FOCUS ON PHYSICS

The Barnes & Noble Focus on Physics titles are prepared under the general editorship of J. Warren Blaker, Associate Professor of Physics, Vassar College.

ABOUT THE AUTHOR

Robert L. Stearns is a graduate of Wesleyan University and holds a Ph.D. from Case Institute of Technology. He has taught at Queens College and is currently Professor and Chairman of the Department of Physics at Vassar College. In addition Professor Stearns holds an appointment as research collaborator with a nuclear physics group at Brookhaven National Laboratory.

College Outline Series

FOCUS ON PHYSICS
Atomic Physics

ROBERT L. STEARNS

Professor of Physics
Vassar College

BARNES & NOBLE, INC., NEW YORK
Publishers *Booksellers* *Since 1873*

L. C. catalogue card number: 73-80000
SBN 389 00016 7

Printed in the United States of America

Preface

FOCUS ON PHYSICS

The ten Focus on Physics volumes in the College Outline Series are concise but comprehensive self-teaching treatments of the most important topics in the first-year physics course.

It has been found that beginning students of physics often require material which can supplement their texts and lectures by supplying, in a somewhat different format, explanations of principles and methods. Such re-enforcement is particularly important in areas where the student is having difficulty. A number of short "outline" books which emphasize the problem-solving aspect of physics are available for this purpose, but we know of no short works which provide a more thorough discussion of underlying principles.

Each Focus on Physics title has been planned to present the subject matter of a particular topic or group of related topics. Each book includes summaries of principles and facts, solved examples, problems with answers, and numerous illustrations.

In order to help the student to attain understanding and mastery of the basic material, these volumes strongly emphasize physical principles. They reflect the detailed approach to physics which is now the standard treatment, and their combined subject matter forms the essence of physical science.

In addition to use by the individual student, the Focus on Physics titles are suitable for class assignment as text supplements. They can also be used for self-instruction by the general reader who is interested in exploring and learning about the elements of modern physics.

J. WARREN BLAKER

General Editor

Table of Contents

Chapter 4 **Atoms With More Than One Electron**

Chapter 5 **The Solid State**

Introduction

ATOMIC PHYSICS

This text covers the fundamental topics appropriate to the introductory study of atomic physics. Because of the supplementary nature of this series, it includes most of the topics treated in the standard texts at this level.

Chapter 1 discusses basic measurements on atoms and electrons such as the determination of q/m for the electron and the measurement of atomic masses. It also introduces definitions and topics used and developed in later chapters, e.g., the atomic mass unit, the electron volt, and a brief summary of special relativity. Chapter 2 offers a standard treatment of the experimental basis for wave-particle duality. Chapter 3 introduces the Bohr model and the quantization of atomic systems and lays the basis for a semi-quantitative discussion of X-rays in Chapter 4. The Mossbauer effect is also briefly touched on in Chapter 3 because of the relatively simple and interesting physics involved and because of its increasing use in other fields such as chemistry. Chapter 5 includes selected topics from the general area of solid state physics, reflecting the increasing coverage of this area at the introductory level.

Chapter 1

Fundamental Measurements on Atoms and Electrons

1-1. Introduction. Although our treatment of atomic physics will be by no means historical, it is instructive to discuss some of the pioneering experiments which have provided us with the information which we have today about the nature of the atom. Thus, in this first chapter we shall discuss how it is possible to actually measure the mass and the electrical charge on the electron despite the fact that this mass is only 9.1×10^{-31} kg. Using similar techniques, we will also discuss how it is possible to measure the masses of the various atoms which are found in nature with accuracies of better than one part in several million. We will also discuss the famous Rutherford scattering experiment which showed how atoms are constructed, providing the first information about the actual distribution of mass and charge in the atom.

1-2. Cathode Rays. As early as 1838, Michael Faraday (1791–1867) found that an electric current flowed when a high voltage (several thousand volts) was applied to metal electrodes at either end of a glass tube from which some of the air had been removed. As the techniques for making pumps improved toward the middle of the nineteenth century, many experiments were done to investigate the properties of the "cathode rays" which appeared to emanate from the negative electrode when the pressure in the cathode ray tube was reduced to about 0.01 mm of mercury. It was observed that these "rays" could be deflected by electric and magnetic fields, and that objects placed in their path produced shadows.

Example 1.1. Calculate the number of molecules per cubic centimeter at atmospheric pressure and at a pressure of 0.01 mm of mercury. (Assume a temperature of 0°C.)

Solution. One can use the fact that at atmospheric pressure and 0°C (STP) one mole of a gas occupies 22,400 cm^3. Since there are 6.02×10^{23} molecules per mole (Avogadro's number), we get $= 6.02 \times 10^{23} \div 22{,}400 = 2.69 \times 10^{19}$ molecules per cubic centimeter. If the temperature and volume stay constant, the pressure will be directly proportional to the number of molecules. Therefore, at 0.01 mm Hg the number of molecules per cubic centimeter should be $2.69 \times 10^{19} \times 0.01 \div 760$, since 760 mm Hg corresponds to atmospheric pressure. This gives $n = 3.54 \times 10^{14}$ molecules per cubic centimeter at 0.01 mm Hg. Although 75,999/76,000 or 99.9987% of the molecules present at atmospheric pressure have been removed, the number of molecules per cubic centimeter is still large. The average distance traveled by a molecule between collisions (the mean free path), however, increases from about 10^{-5} cm to almost 1 cm.

Considerable controversy developed during the latter part of the nineteenth century as to whether the cathode rays were particles or waves. We will describe one of the many experiments performed during this period to illustrate how it was possible to measure the ratio of charge to mass of the electron (which was the name given to the "particles" which were found to constitute the cathode rays). This experiment was performed by the English scientist J. J. Thomson (1856–1940) in 1897 and involved the use of both electric and magnetic fields. When this same charge-to-mass ratio was observed for many different cathode materials in cathode ray tubes (as well as for the "particles" emitted by a hot filament and for so-called photoelectrons emitted when light was incident on a metal plate), it became clear that the electron was an important and fundamental part of all matter.

1-3. Thomson's Measurement of q/m. Figure 1.1 illustrates schematically the experimental arrangement used by Thomson in measuring the ratio of charge to mass for the electron. The elec

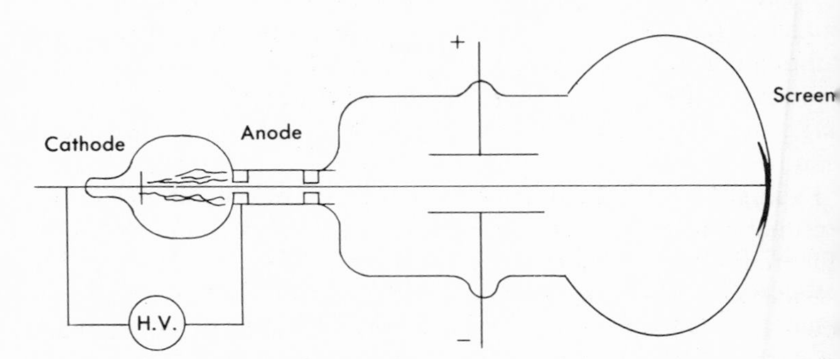

Figure 1.1. Schematic drawing of the equipment used by J. J. Thomson to measure q/m for the electron.

trons were allowed to pass through a slit in the anode and were further collimated by a second slit. The beam of electrons then passed between two parallel metal plates and on to a fluorescent screen which indicated the position of the beam.

If an electric field is applied between the two plates, the beam will be bent upward if the upper plate is positive. If a magnetic field is produced in the same region using appropriately positioned external coils, the beam will be bent down if the field is directed into the paper. If one assumes that the electric and magnetic fields are uniform over, and limited to, the region between the plates (which can only be approximately true), the beam can be adjusted for zero deflection—in which case the electric force and the magnetic force on the electrons must be equal. Therefore:

$$\mathbf{E}q = q(\mathbf{v} \times \mathbf{B}) \tag{1.1}$$

where **E** is the electric field between the plates, q the charge on the electron, **B** the magnetic flux density, and **v** the velocity of the electrons. Equation 1.1 gives us: (if **v** is $\perp$ to **B**)

$$v = \frac{E}{B} \tag{1.2}$$

and permits the determination of q/m if the radius of curvature of the electrons in the magnetic field can be measured. A charged particle in a uniform magnetic field moves in a circular path whose radius is given by:

$$\frac{mv^2}{r} = Bqv \tag{1.3}$$

If equation 1.3 is combined with equation 1.2, we obtain for the ratio of charge to mass:

$$\frac{q}{m} = \frac{v}{Br} = \frac{E}{B^2 r} \tag{1.4}$$

Example 1.2. Calculate q/m for the electron if, in an experiment similar to that described above, the potential difference (V) between the parallel plates is 200 volts, the plate separation is 1 cm, the magnetic field required for no deflection is 10^{-3} webers/m^2, and the radius of curvature in the magnetic field alone is 11.4 cm.

Solution. Using equation 1.4 directly, remembering that E =

V/d, and working in the MKS system we get:

$$\frac{q}{m} = \frac{E}{B^2 r} = \frac{V}{dB^2 r}$$

$$= \frac{200}{0.01 \times 10^{-6} \times 0.114} = 1.76 \times 10^{11} \text{ coul/kg}$$

Example 1.3. Calculate the velocity of the electrons in this experiment.

Solution. We use equation 1.2 directly and get: ($E = 2 \times 10^4$ v/m)

$$v = \frac{E}{B} = 2 \times \frac{10^4}{10^{-3}} = 2 \times 10^7 \text{ m/sec}$$

Example 1.4. In the problem above, through what potential difference must the electrons have moved in the cathode ray tube before passing through the parallel plates?

Solution. The increase in kinetic energy of the electrons must equal the change in electrical potential energy so that:

$$\tfrac{1}{2} mv^2 = Vq$$

$$V = \frac{\frac{1}{2} mv^2}{q} = \frac{\frac{1}{2} v^2}{q/m}$$

$$= \frac{\frac{1}{2} \times (2 \times 10^7)^2}{1.76 \times 10^{11}}$$

$$= 1136 \text{ volts}$$

Example 1.5. Consider two parallel plates 20 cm long, separated by 4 cm. A potential difference of 10 volts is maintained between the plates. An electron moving with a velocity of 10 m/sec parallel to the plates enters the region between the plates halfway between them. Calculate the displacement of the electron from a plane midway between the plates when it reaches the far end of the plates.

Solution. The displacement must be given by $s = \frac{1}{2} at^2$ where a is the acceleration experienced by the electron due to the electric field between the plates and t is the time necessary for the electron to travel 20 cm. The electric field between the plates is $E = V/d = 10 \div 0.04 = 2.5 \times 10^2$. The acceleration must be given by $a = F/m = Eq/m = 2.5 \times 10^2 \times 1.6 \times 10^{-19} \div 9.1 \times$

10^{-31}, and $a = 0.439 \times 10^{14}$ m/sec^2. The time is simply $t = l/v = 0.20 \div 10^7 = 2 \times 10^{-8}$ sec.

The displacement is therefore:

$$s = \frac{1}{2}at^2 = \frac{1}{2} \times 0.439 \times 10^{14} \times 4 \times 10^{-16}$$
$$= 0.878 \times 10^{-2}\,\text{m}$$
$$= 0.878\,\text{cm}$$

Example 1.6. Calculate the magnitude and direction of the magnetic field required to permit an electron having a velocity of 2×10^7 m/sec to move undeflected between two parallel plates, separated by 5 cm, between which a potential difference of 1000 volts is maintained (see Figure 1.2).

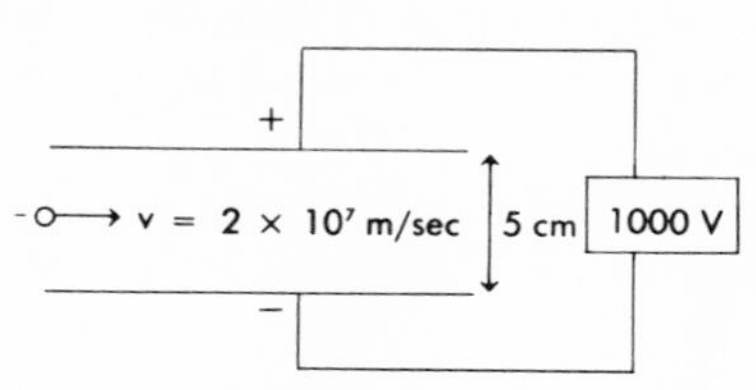

Figure 1.2.

Solution. The relationship between E and B is given by equation 1.2 as $v = E/B = 2 \times 10^{-7}$. The electric field between the plates is $E = V/d = 1000 \div 0.05 = 2 \times 10^4$ volts/meter. Thus:

$$B = \frac{E}{v} = \frac{2 \times 10^4}{2 \times 10^{-7}}$$ $= 10^{-3}$ webers/m^2 = 10 gauss perpendicular to v in Figure 1.2 and directed into the page.

1-4. Measurement of the Electronic Charge. Many experiments were done in the 1890's, primarily in England and France, to measure the fundamental electric charge found in nature. It was usually assumed that this was probably the charge on the electron and that all charges found in nature might be integral multiples of this fundamental charge. The early experiments were done utilizing the fact that in saturated or nearly saturated water vapor, droplets tended to form on electrically charged ions. The droplets were collected and the charge on each droplet estimated by measuring the total charge collected and dividing by the number of drops. The number of drops was estimated by dividing the total mass of water collected by the mass of each drop as estimated by observing the rate of fall of the drops under the influence of gravity, from which the radius could be determined as in the Millikan experiment described below. If it is assumed that each drop is singly charged, then the value obtained would be that of the fundamental charge or the charge on the electron. Using this technique, values of about 10^{-19} coulombs were obtained. These values were, in retrospect, surprisingly close to the currently accepted value of 1.6×10^{-19} coulombs.

A much more definitive experiment which strongly indicated that the charge on the electron was the fundamental unit of electric charge in nature was performed by the American physicist Robert A. Millikan (1868–1953) at the University of Chicago in the years between 1907 and 1913.

Millikan made measurements on charged oil drops as they moved in the uniform electric field between parallel metal plates. A schematic diagram of his apparatus is shown in Figure 1.3. The

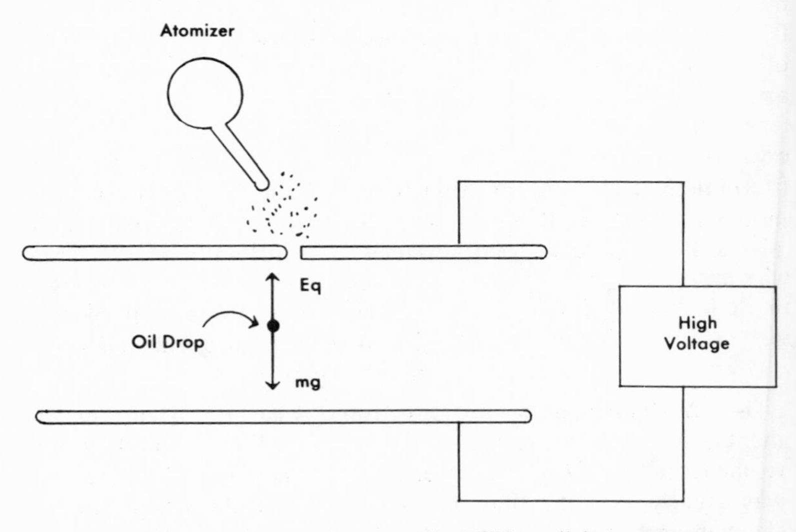

Figure 1.3. A schematic diagram of the Millikan oil drop apparatus.

oil drops, produced by a simple spray gun or atomizer, were illuminated and viewed from the side using a telescope. The individual drops, which appear as bright spots in the field of the telescope, can be followed as they fall under the force of gravity alone or are made to move upward under the influence of the electric field between the plates. Perhaps the most obvious and simple procedure is to balance the downward gravitational force on the drop with the upward electrical force. In such a case:

$$Eq = mg = \tfrac{4}{3}\pi r^3 (\rho - \rho_a)g \qquad (1.5)$$

which assumes that the drops are spheres of radius r. In this equation, ρ is the density of the oil, and ρ_a is the density of the air, which is included to account for the buoyant force exerted by the air. Solving for the charge on the drop, we get:

$$q = \frac{\tfrac{4}{3}\pi r^3 (\rho - \rho_a)g}{E} \qquad (1.6)$$

The only quantity in the equation which is unknown is the radius, r. Since the radius is too small to measure directly, other means must be used. The technique applied by Millikan was to allow the drops to fall freely under the influence of gravity and measure their terminal velocity, i.e., the velocity at which the frictional force experienced by the drops is equal to mg, the force of gravity. (This is quite analogous to the terminal velocity acquired by a parachutist. The larger the chute, the smaller the terminal velocity.) A relationship known as Stokes' law was well established at the time. It related the terminal velocity and the frictional force as follows:

$$f = Krv_t \qquad (1.7)$$

where K is a constant of proportionality. By equating this force to the force of gravity, we obtain an expression relating the terminal velocity and the radius of the drop.

$$\tfrac{4}{3}\pi r^3 (\rho - \rho_a)g = Krv_t$$

so that:

$$v_t = \frac{4}{3K}\pi r^2 (\rho - \rho_a)g \qquad (1.8)$$

which indicates that the terminal velocity is proportional to the cross sectional area of the drop.

Using this expression to obtain the radius in terms of v_t, i.e., by combining equations 1.8 and 1.6 we get:

$$q = \frac{\left(\frac{3}{4\pi(\rho - \rho_a)g}\right)^{\frac{1}{2}} (Kv_t)^{\frac{3}{2}}}{E} \qquad (1.9)$$

Although Millikan's early experiments were carried out in this manner, it was found that there were practical difficulties in deciding when the drops were exactly balanced. To eliminate this problem, Millikan measured the terminal velocity in the upward direction when the electric field was increased so as to produce a force on the charged drop greater than the force of gravity. Under these conditions the drop is in equilibrium, i.e., moving with constant velocity in the vertical direction when:

$$\mathrm{Eq} - K\mathrm{rv}_t' = \tfrac{4}{3}\pi \mathrm{r}^3 (\rho - \rho_a)\mathrm{g} \tag{1.10}$$

where v_t' represents the upward terminal velocity.

Using equations 1.8 and 1.10 to eliminate r, one again can get an expression for the charge on the drop in terms of the two terminal velocities and the electric field between the plates. This expression is:

$$\mathrm{q} = \left(\frac{3}{4\pi(\rho - \rho_a)\mathrm{g}}\right)^{\frac{1}{2}} \left(\frac{K^{\frac{3}{2}}}{\mathrm{E}}\right)\left(\mathrm{v}_t' + \mathrm{v}_t\right)\left(\mathrm{v}_t\right)^{\frac{1}{2}} \tag{1.11}$$

It was often possible to follow the same drop for long periods of time. During this time, the charge on the drop often changed and in fact could be made to change by exposing the apparatus to ionizing *X* rays. Such changes usually involved only the addition of a few atoms to the drop and hence produced no measurable change in mass. Note the simplicity which results if one uses the same drop and measures v_t' for two different values of the charge. In equation 1.11, v_t' is the only quantity which changes and one obtains:

$$\mathrm{q}_1 - \mathrm{q}_2 = \text{constant} \times (\mathrm{v}_{t_1}' - \mathrm{v}_{t_2}') \tag{1.12}$$

The constant in equation 1.12 is constant *only* if the same drop is used and if the electric field is not changed. Thus it was often easiest to follow the changes in the charge on one drop.

Millikan measured many charges on oil drops and never observed a charge which was not a multiple of the charge usually designated by $e = 1.6 \times 10^{-19}$ coulombs. This fact, of course, does not prove that other "non-integral" charges do not exist in nature, but over the years Millikan's experiment has been generally credited with showing this because it has been supported so consistently by many other experiments involving measurement of the fundamental charge on the electron.

Example 1.7. The plates in an oil drop apparatus are separated by a distance of 0.5 cm. If a drop of mass 5×10^{-15} kg with a single electronic charge is to be balanced between the plates, what voltage must be applied?

Solution. The upward electrical force must equal the force of gravity on the oil drop. Therefore:

$$Eq = \frac{V}{d} q = mg$$

$$V = \frac{mgd}{q} = \frac{5 \times 10^{-15} \times 9.8 \times 0.5 \times 10^{-2}}{1.6 \times 10^{-19}}$$

$$= 1530 \text{ volts}$$

Example 1.8. An oil drop apparatus with a spacing between the plates of 1 cm is used to balance a drop of mass 1.63×10^{-14} kg. The potential difference between the plates is 1000 volts. If the charge changes so that a voltage of 1110 volts is required for balance, how many charges has the drop gained or lost?

Solution. For balance we must have mg = nEq, where n is the number of charges and E is the electric field which must be given by $E = V/d = 1000 \div 0.01 = 10^5$ volts per meter.

When 1000 volts provides the balance:

$$n = \frac{mg}{Eq} = \frac{1.63 \times 10^{-14} \times 9.8}{10^5 \times 1.6 \times 10^{-19}}$$

$$= 10 \text{ charges}$$

When 1110 volts provides the balance:

$$n = \frac{mg}{Eq} = \frac{1.63 \times 10^{-14} \times 9.8}{1.11 \times 10^5 \times 1.6 \times 10^{-19}}$$

$$= 9 \text{ charges}$$

Therefore the drop has lost one charge.

Example 1.9. Calculate the terminal velocity v_t for an oil drop of radius 10^{-6} m falling freely under the influence of gravity. The constant K is given as 3.45×10^{-4} kilograms per meter per second. Calculate also the terminal velocity for a rain drop of radius 5×10^{-4} m. Assume that the density of each is 1000 kg/m^3. (Neglect the density of air.)

Solution. From equation 1.8:

$$v_t = \frac{4}{3K} \pi r^2 \rho g$$

For the oil drop:

$$v_t = \frac{4}{3 \times 3.45 \times 10^{-4}} \times 3.14 \times 10^{-12} \times 10^{3} \times 9.8$$
$$= 1.19 \times 10^{-4}\ \text{m/sec} = 0.0119\ \text{cm/sec}$$

For the rain drop, all the numbers will be the same except for the radius. Thus the terminal velocity will be given by multiplying the above value by the ratio of the squares of the radii.

$$v_t = 0.0119\ \text{cm/sec} \times \frac{(5 \times 10^{-4})^2}{(10^{-6})^2}$$
$$= 29.8 \times 10^{2}\ \text{cm/sec} = 29.8\ \text{m/sec} \approx 67\ \text{mi/hr}$$

Example 1.10. Calculate the upward terminal velocity of an oil drop of mass 10^{-14} kg and a charge of 5 electronic charges if the electric field between the plates of the oil drop apparatus has a value of 3.82×10^{5} volts/m. The radius of the drop is 1.35×10^{-6} m.

Solution. The terminal velocity is reached when the frictional force is equal to the difference between the upward electrical force and the force of gravity, i.e., $Krv_t' = Eq - mg$ so that:

$$v_t' = \frac{Eq - mg}{Kr}$$

$$v_t' = \frac{(3.82 \times 10^{5} \times 5 \times 1.6 \times 10^{-19}) - (10^{-14} \times 9.8)}{3.45 \times 10^{-4} \times 1.35 \times 10^{-6}}$$
$$= (30.5 - 9.8) \times \frac{10^{-14}}{4.66 \times 10^{-10}}$$
$$= 4.45 \times 10^{-4}\ \text{m/sec} = 0.0445\ \text{cm/sec}$$

1-5. The Nuclear Atom. With the identification of the negatively charged electron and some knowledge of its charge and mass it became clear that, since matter was electrically neutral, most of the mass associated with atoms must be connected with a positive charge. Just how the positive and negative charges might be arranged in atoms was not at all clear however. During the early 1900's, one of the more popular models of the atom was proposed by J. J. Thomson. It was often referred to as the "plum pudding" model and pictured the atom as a fluid sphere of heavy positive charge through which were distributed the negatively charged electrons, much as raisins in a plum pudding, to produce electrical neutrality. The electrons could be caused to vibrate about equilibrium positions as the result of thermal or electrical disturbances, and Thomson calculated that their frequencies should be about the same as the frequencies associated with visible light. He could not, however, provide any mechanism to explain the experimentally observed fact that light from gas discharges, when analyzed with prisms and gratings, was made up of discrete frequencies and was not continuous.

The experiment which showed the "plum pudding" model to be incorrect and provided us for the first time with the model of the atom which, with refinements, is still in use today was performed in 1911 in the laboratory of Ernest Rutherford (1871–1937) at the University of Manchester in England.

At that time it was known that certain radioactive materials, such as radium, gave off very heavy, very rapidly moving particles called alpha particles. The ratio of charge to mass had been measured, as had the charge, and it was known that the alpha particles had a positive charge twice as large as the charge on the electron and a mass about four times that of the hydrogen atom or about 6.68×10^{-27} kg, which is about 7300 times the mass of the electron. The velocities of these alpha particles from natural radioactive substances range from 1.2×10^7 m/sec to 2×10^7 m/sec.

A drawing of one of the actual experimental arrangements used in Rutherford's laboratory is shown in Figure 1.4. The airtight box B was evacuated through the tube T since the alpha particles have a range of only a few centimeters at atmospheric pressure in air. Alpha particles striking the scintillating zinc sulphide screen S (fastened to the telescope) produced flashes which were observed visually. The box, telescope, and screen could be rotated while the source R and foil F remained fixed.

Rutherford, or more exactly one of his students named Marsden, measured the relative number of alpha particles scattered at angles up to 150° from a thin gold foil F about 4×10^{-5} cm thick. Thomson's model of the atom predicted that all the alpha particles would go straight through the foil experiencing only very slight deflections. Rutherford was amazed when he learned that a very few of the alpha particles were scattered at large angles, some coming almost directly backwards. To Rutherford, this was like shooting a cannon at a piece of tissue paper and having the shell bounce off in the direction from which it came.

Since it was virtually impossible that such large-angle scatter-

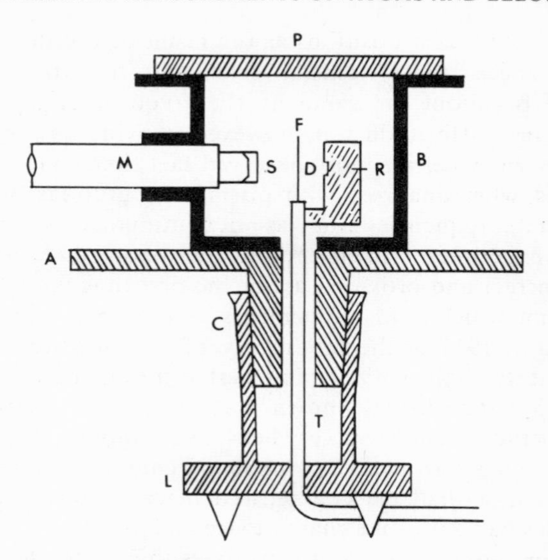

Figure 1.4*a*. The apparatus* used by Rutherford in his alpha particle scattering experiment which first revealed the nuclear atom. (From Geiger and Marsden, *Phil. Mag.*, 1913, Vol. 25, p. 604.)

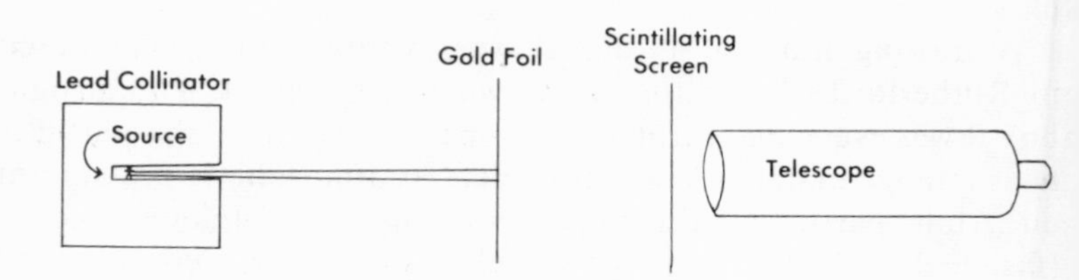

Figure 1.4*b*. Schematic diagram of the Rutherford experiment.

ings could result from successive encounters with electrons, Rutherford concluded that the scatterings must have resulted from single encounters with the small, massive, and probably positively charged "nuclei" of the gold atoms in the foil. He estimated the radius of this nucleus by calculating the distance of closest approach experienced by an alpha particle in a head-on collision with a positively charged gold nucleus. This is done by equating the original kinetic energy of the alpha particle to the

*This equipment was used by Geiger and Marsden in somewhat later experiments performed to test the predictions of Rutherford's theory of the nuclear atom.

electric potential energy of the system at the point of closest approach and solving for the center-to-center separation of the two charged particles.

Rutherford obtained values of this distance of closest approach which were on the order of 10^{-12} cm. Although these values were almost certainly larger than the "nuclear radius," they were so small that it became clear, after additional experiments were done to test the predictions of Rutherford's atomic model, that the nucleus was almost inconceivably smaller than the atom for which it provided more than 99.9% of the mass. It was known that atomic radii were roughly 10^{-8} cm, so it is easy to see why most of Rutherford's alpha particles went undeflected through the gold foil, which was about 1000 atomic diameters thick. A feeling for the relative size of things in the atom is provided by realizing that if the nucleus were roughly the size of a basketball, one foot in diameter, the corresponding atomic diameter would be about two miles.

Example 1.11. Calculate the distance of closest approach for an alpha particle, of mass 6.68×10^{-27} kg and velocity 2×10^{7} m/sec, incident on a gold nucleus in a head-on collision. Assume the gold nucleus to be stationary (actually it would recoil) and to have a charge of 79 electronic charges.

Solution. We equate the original kinetic energy of the alpha particle to the potential energy of the two particles at their distance of closest approach, d.

$$\frac{1}{2}mv^2 = \left(\frac{1}{4\pi\epsilon_0}\right)\left(\frac{q_\alpha Q}{d}\right)$$

where q_α is the charge on the alpha particle $= 2 \times 1.6 \times 10^{-19}$ coul, and the charge on the gold nucleus is $Q = 79 \times 1.6 \times 10^{-19}$ coul. Recall that $1/4\pi\epsilon_0 = 9 \times 10^{9}$ newton-m^2/coul^2. Thus:

$$d = \frac{2q_\alpha Q}{4\pi\epsilon_0 mv^2}$$

$$d = 18 \times 10^9 \times 2 \times 79 \times \frac{(1.6 \times 10^{-19})^2}{(6.68 \times 10^{-27} \times 4 \times 10^{14})}$$

$$d = 2.72 \times 10^{-14}\,\text{m} \approx 3 \times 10^{-12}\,\text{cm}$$

Therefore, the nuclear model of the atom first proposed by Rutherford pictures the atom with a very small, massive, positively charged nucleus about which the negatively charged electrons move in some way. Rutherford's model stimulated further calculations—particularly those concerned with the light emitted by hydrogen atoms. We shall discuss these results in Chapter 2.

We now know that the nucleus may be thought of as containing two kinds of particles called the neutron and proton. The neutron is uncharged and slightly heavier than the proton, being 1838.6 times more massive than the electron. The proton has a positive charge equal in magnitude to that on the electron and is 1836.1 times heavier than the electron. For the lighter atoms found in nature, the number of neutrons tends to equal the number of protons; but as the atoms get larger and heavier, more neutrons than protons are found in the nucleus. In the case of uranium, for example, there are 92 protons and 146 neutrons.

The number of protons and electrons in an atom are the same since the atom is electrically neutral. This number is called the atomic number. The number of neutrons plus protons is called the mass number and is a rough measure of the relative masses of the various atomic species.

1-6. The Electron Volt. In the fields of atomic and nuclear physics it is customary to specify energies in terms of a unit called the electron volt. A particle is said to have an energy of one electron volt if it is singly charged and falls through a potential difference of one volt. If the particle had a charge equal to twice that on the electron and was accelerated through a potential difference of one volt, its energy would be two electron volts (ev). We can calculate the relationship between the joule and the electron volt by calculating the energy, in joules, of a singly-charged particle which is allowed to move through a potential difference of one volt. Thus:

$$E = Vq = 1 \times 1.6 \times 10^{-19} \text{ joules}$$

or:

$$1 \text{ electron volt} = 1.6 \times 10^{-19} \text{ joules} \qquad (1.13)$$

Often energies are expressed in millions of electron volts, since, as indicated by equation 1.13, the electron volt is a rather small unit of energy. The abbreviation Mev is used for 10^6 electron volts, and the term Bev or Gev is used for 10^9 ev.

Example 1.12. A proton and an alpha particle are accelerated from rest through a potential difference of 100 volts. Calculate the energy of each in electron volts and the velocity of each.

Solution. From the definition of the electron volt, the energy of the proton is 100 ev while the energy of the alpha particle is 200 ev since it is doubly charged.

Since $\frac{1}{2}mv^2 = Vq$, we have $v = \sqrt{2Vq/m}$. For the proton, $q = 1.6 \times 10^{-19}$ and $M = 1.67 \times 10^{-27}$ kg.

$$v = \sqrt{\frac{2 \times 100 \times 1.6 \times 10^{-19}}{1.67 \times 10^{-27}}}$$

$$= \sqrt{191.6 \times 10^{8}} = 13.84 \times 10^{4} \text{ m/sec}$$

For the alpha particle, we note that the charge is twice as large and the mass four times as large so that the velocity must be less by a factor of $\sqrt{2/4} = 0.707$ and $v = 9.78 \times 10^4$ m/sec.

Example 1.13. Calculate the energy, in electron volts, of a 20 g marble dropped from a height of 1 m.

Solution. The kinetic energy of the marble as it hits the floor is equal to its change in potential energy, mgh. Here $m = 0.02$ kg and $h = 1$ meter.

$$mgh = 0.02 \times 9.8 \times 1.0 = 0.196 \text{ joules}$$

$$mgh = \frac{0.196}{1.6 \times 10^{-19}} = 0.122 \times 10^{19} \text{ ev}$$

Example 1.14. Consider a proton and an electron (at rest) separated by a distance of 0.528×10^{-10} m. Calculate the energy required to move the electron a large distance (i.e., to infinity) from the proton.

Solution. The potential energy of the system is given by $V = (1/4\pi\epsilon_0)(q_1 q_2/r)$, and, since at infinity this is zero, the energy required is simply $(1/4\pi\epsilon_0)(q^2/r)$, where $q = 1.6 \times 10^{-19}$ coulombs.

$$V = \frac{9 \times 10^9 \times (1.6 \times 10^{-19})^2}{0.528 \times 10^{-10}}$$

$$= \frac{23.04 \times 10^{-29}}{0.528 \times 10^{-10}} = 43.64 \times 10^{-19} \text{ joules}$$

$$= \frac{43.64 \times 10^{-19} \text{ joules}}{1.6 \times 10^{-19} \text{ joules/ev}} = 27.25 \text{ ev}$$

Example 1.15. Calculate the magnitude of the magnetic field required to bend 1000 ev protons in a circular path of radius 1 m.

Solution. Equating the magnetic force to the required centripetal force we get $Bqv = mv^2/r$ so that:

$$B = \frac{mv}{qr}$$

Since $Vq = \frac{1}{2}mv^2$, then $v = \sqrt{2Vq/m}$. Substitution then gives:

$$B = \left(\frac{2\,Vq}{m}\right)^{\frac{1}{2}}\left(\frac{m}{qr}\right) = \frac{\sqrt{2\,Vqm}}{qr}$$

$$= \frac{\sqrt{2 \times 1000 \times 1.6 \times 10^{-19} \times 1.67 \times 10^{-27}}}{1.6 \times 10^{-19} \times 1}$$

$$= 4.57 \times 10^{-3}\ \text{webers/m}^2$$

1-7. Some Facts from the Special Theory of Relativity. Although it is beyond the scope of this book to discuss in detail the Special Theory of Relativity, it is important for us to discuss some of the results of this theory which are significant in the general area of atomic physics.

In 1887 the American scientists A. A. Michelson and E. W. Morley performed an experiment which bears their name and which showed that the speed of light, as measured by any observer, is independent of the speed of the source and the observer. This fact is the basis for the Special Theory of Relativity which revolutionized the field of physics and was proposed by Albert Einstein (1879–1955) in 1905.

This fundamental postulate was very hard for many physicists to swallow early in the twentieth century because it predicted things that were contrary to what might be called "common sense" based on our everyday experience. For example, if two automobiles were to approach one another, one at 60 mi/hr relative to the ground and one at 80 mi/hr relative to the ground, one would predict that a person riding in one would measure their relative velocity as 140 mi/hr. The Special Theory of Relativity claims that this isn't quite so and becomes less and less the case as the velocities of the automobiles approach the velocity of light. It predicts that the relative velocity is given by:

$$v_R = \frac{v_1 + v_2}{1 + v_1 v_2/c^2} \tag{1.14}$$

where c is the velocity of light.

It is of course ridiculous to apply this to automobiles and indeed the relativistic effect is completely negligible at normal automotive velocities, but if we imagine that two spacecraft somehow approach each other at speeds of $0.8c$ and $0.6c$, we might predict that the occupants of each would see the other approach at $1.4c$. Relativity says, however, that according to equation 1.14 the measured velocity would be $(1.4/1.48)c$ or $0.946c$. Note that if v_1 or v_2, or both, are set equal to c, then $v_R = c$. Thus, special relativity predicts that $c = 3 \times 10^8$ m/sec is the upper "speed limit" found in nature. Despite the apparent strangeness of this result and others predicted by special relativity, so many of its predictions have been verified by experiment that little if any doubt remains as to the validity of this theory.

One of the predictions of the Special Theory was that moving clocks should appear to run more slowly when observed in motion relative to the observer than when at rest. In order for this effect to be of any significance, this velocity should be an appreciable fraction of the velocity of light. We can derive rather simply the factor by which moving clocks appear to slow down. A similar factor appears in some of the other expressions which we will not attempt to derive, and this brief discussion will at least provide some feeling for relativistic effects.

Let us consider a rather special kind of clock as indicated in Figure 1.5. This clock consists of a light source which sends a pulse of light upward to the mirror *M*. The light is reflected back to a detector near the source which causes a second pulse to be emitted by the source. We can define one "tick" of this clock as the time required for the light pulse to traverse the distance d. If

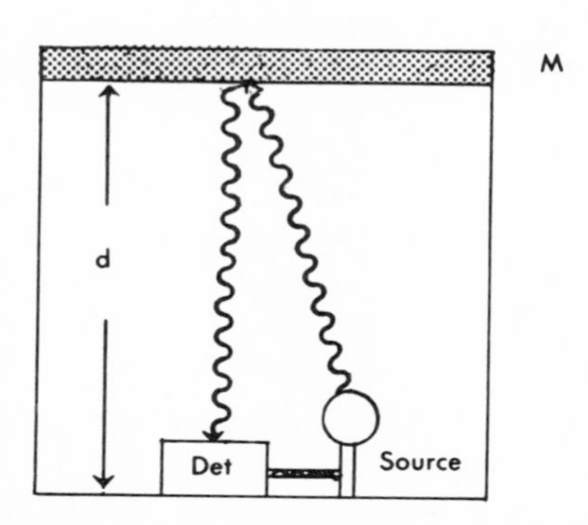

Figure 1.5*a*. A clock for use in thought experiment involving time dilation.

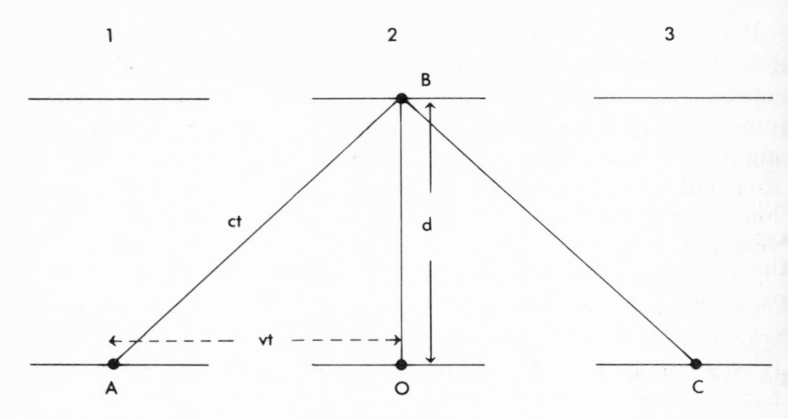

Figure 1.5*b*. Three positions of a clock moving with velocity v which is an appreciable fraction of c, the velocity of light.

the "clock" is moving to the right with a velocity v which is an appreciable fraction of the velocity of light, the light will appear to a stationary observer to move along the diagonal path ABC as indicated in Figure 1.5*b*. In position 1, the light leaves the source; at position 2, it reaches the mirror; and at position 3, it arrives back at the detector next to the source. The time required for the light to travel the distance d between the source and the mirror in a stationary clock is simply $t_0 = d/c$ where c is the velocity of light. However, if we require that the velocity of light always be c, no matter how the source or observer move, we see from the triangle ABC that $c^2t^2 = v^2t^2 + d^2$, where v is the velocity of the clock and t is the time required for the clock to move from position 1 to position 2 and also the time for one "tick" of our clock. From the above equation, $t^2 = d^2/(c^2 - v^2)$ and dividing top and bottom by c^2 gives $t^2 = (d^2/c^2)/(1 - v^2/c^2)$ so that:

$$t = \frac{d/c}{\sqrt{1 - v^2/c^2}}$$

If $t_0 = d/c$ we get:

$$t = \frac{t_0}{\sqrt{1 - v^2/c^2}} \qquad (1.15)$$

and the moving clock appears to run slowly by a factor of $1/\sqrt{1 - v^2/c^2}$.

This relativistic time dilation is virtually never encountered in our everyday experience, but physicists working with high energy particles must deal with it constantly. A good example is provided by the π-meson, which can be produced using high energy accelerators. It has a mass about 270 times that of the electron and is unstable, decaying into a positive μ-meson and a neutrino with a half-life ($t_{\frac{1}{2}}$) of 1.8×10^{-8} sec. This means that if there were no relativistic time dilation, a beam of π-mesons coming from a target in an accelerator would be reduced in intensity (assuming the beam to be collimated or focused so that the particles move parallel to one another) by a factor of two in a distance given by $v_\pi t_{\frac{1}{2}}$. For π-mesons with a velocity of $0.954c$, this distance would be 5.15 meters. At this velocity, the factor $1/\sqrt{1 - v^2/c^2}$ is:

$$\frac{1}{\sqrt{1 - (0.954c)^2/c^2}} = \frac{1}{\sqrt{1 - 0.91}} = 3.33$$

so that relativity predicts that the half-life at this velocity will be 3.33 times as large and that the distance traveled by the beam will be 3.33×5.15 m $= 17.15$ m before it is reduced in intensity by a factor of two due to decay of the mesons. Experiments of this kind have been done, and agreement is always obtained within the accuracy of the experimental equipment.

Example 1.16. Calculate the factor by which the half-life of the π-meson is increased if its velocity is $0.8c$.

Solution. Since $t = t_0/\sqrt{1 - v^2/c^2}$, we must calculate $1/\sqrt{1 - v^2/c^2} = 1/\sqrt{1 - 0.64} = 1/\sqrt{.36} = 1.67$, which is the factor by which the meson half-life is increased.

An interesting consequence of the relativistic time dilation concerns the fact that all processes from the ticking of clocks to the natural frequencies characteristic of atoms should experience this slowing down. Thus, scientists are confident that all biological processes, and therefore the aging process, should also be affected. This has led to considerable discussion of what is often called the "twin paradox." This so-called paradox concerns identical twins. One twin goes off to a distant part of our galaxy at a speed close enough to the velocity of light so that the twin remaining on the earth determines that the traveling clocks run significantly more slowly than those on earth—perhaps at one-thirtieth the normal rate. If the first twin goes out into space for 15 years according to earth clocks and then turns around and returns at the same speed, his journey should take 30 years according to his twin who remains behind. Thus, if the twins were

30 years old at the start of the journey, the earthbound twin should be 60 years old when he returns; but the traveling twin, if his clocks indeed run at one-thirtieth the normal rate, should be only 31 years old. Of course, this change in the rate of his clocks is not apparent to the traveling twin.

One might complain that from the point of view of the traveling twin, the earthbound twin appears in motion and hence his clock should also appear to run slowly when observed by the traveling twin. This, of course, is true but if one carefully defines such terms as "observer" and "rate" of a clock and considers the asymmetry introduced by the fact that the traveling twin must stop and reverse his direction and therefore change reference systems, it can be shown[1] that indeed the traveling twin should be expected to age, in our example, at only one thirtieth the rate of his brother on earth.

Experiments such as described above, in which human aging is drastically affected, will probably not be done in the foreseeable future, if ever, because of the tremendous energy requirements. In our example, sufficient energy would have been required to increase the mass of the space ship by a factor of thirty. If, somewhat arbitrarily, one assumes the ship to have a mass of 1000 kg (about a ton), the quantity $mc^2 = 30 \times 1000 \times 9 \times 10^{16} = 2.7 \times 10^{21}$ joules (see equations 1.15′ and 1.17). This is somewhat more than the estimated total[2] energy required per year in 1968 by the whole world from water power, coal, oil, and natural gas. Thus, the energy requirements, even at substantially smaller velocities, are staggering.

A second prediction of Special Relativity is that the masses of moving particles must increase according to the relationship:

$$m = \frac{m_0}{(1 - v^2/c^2)^{\frac{1}{2}}} \qquad (1.15')$$

where m_0 is the mass at zero velocity or the "rest mass." This, too, is easily verified by experiment and has been checked and used hundreds of times in experiments involving the deflection of rapidly moving charged particles in electric and magnetic fields.

[1]For an excellent discussion of all aspects of relativity, see *Spacetime Physics* by Edwin Taylor and John A. Wheeler, W.H. Freeman and Co., San Francisco.

[2]*Energy Does Matter* by W. Emmerich, M. Gottlieb, C. Helstrom and W. Stewart, p. 200, A Westinghouse Search Book, Walker and Co., New York.

Related to equation 1.15 is another prediction concerning the energy of rapidly moving particles. This is the familiar $E = mc^2$ relating mass and energy and indicating that matter is just another form of energy. Thus, we may think of the total energy of a moving particle as being:

$$E_T = mc^2 = K.E. + m_0c^2 \qquad (1.16)$$

where the latter term is the so-called rest energy or the energy equivalent of the "rest mass" of the particle. The kinetic energy of a moving particle is then:

$$K.E. = mc^2 - m_0c^2 \qquad (1.17)$$

where m is, of course, given by equation 1.15.

Example 1.17. Calculate the energy equivalent in Mev of the rest mass of the electron and proton.

Solution. Using $E = mc^2$ and the fact that the mass of the electron is 9.11×10^{-31} kg and c, the velocity of light, is 3×10^8 m/sec:

$$E = 9.11 \times 10^{-31} \times 9 \times 10^{16} = 81.99 \times 10^{-15} \text{ joules}$$

$$= \frac{81.99 \times 10^{-15}}{1.6 \times 10^{-19}} \text{ ev}$$

$$= 0.511 \times 10^6 \text{ ev} = 0.511 \text{ Mev}$$

For the proton, $m = 1.67 \times 10^{-27}$ kg so that:

$$E = mc^2 = 1.67 \times 10^{-27} \times 9 \times 10^{16} = 15.03 \times 10^{-11} \text{ joules}$$

$$= 15.03 \times \frac{10^{-11}}{1.6 \times 10^{-19}} = 9.38 \times 10^8 \text{ ev} = 938 \text{ Mev}$$

Example 1.18. What velocity must a spaceship have with respect to an observer so that the mass of the spaceship, as determined by this observer, is increased by a factor of thirty?

Solution. In order for the mass to be increased by a factor of 30, the spaceship must have a velocity such that $1/\sqrt{1 - v^2/c^2} = 30$. For this to be the case, we must have: $v^2/c^2 = 1 - 1/900 = 1 - 0.001111 = 0.99889$, so that $v = 0.9994c$.

Example 1.19. Calculate the factor by which the masses of the electron, proton, and alpha particle increase when each has an energy of 20 Mev.

Solution. This is perhaps most easily done by utilizing equation 1.16 and taking the ratio of the total energy to the rest energy. For the electron:

$$\frac{m}{m_0} = \frac{mc^2}{m_0c^2} = \frac{\text{K.E.} + m_0c^2}{m_0c^2}$$

$$= \frac{20 + 0.511}{0.511} = 40.1$$

For the proton:

$$\frac{m}{m_0} = \frac{20 + 938}{938} = 1.0213$$

For the alpha particle ($m = 6.65 \times 10^{-27}$ kg):

$$mc^2 = 3730 \text{ Mev}$$

$$\frac{m}{m_0} = \frac{20 + 3730}{3730} = 1.0054$$

Using the fact that the momentum of a particle is always given by mv, where $m = m_0/\sqrt{1 - v^2/c^2}$, and the fact that the total energy is $E_T = mc^2$ one obtains by eliminating v and m from these equations:

$$E_T^2 = p^2c^2 + (m_0c^2)^2 \tag{1.18}$$

For relativistic particles, i.e., those with velocities which are an appreciable fraction of the velocity of light, one must use p as calculated from equation 1.18, if the energy is known, or simply from p = mv. This is often important in calculating the radius of curvature of particles in magnetic fields.

Often p is used in units of Mev/c in equation 1.18 so that all three terms have the units Mev^2. To convert p in units of Mev/c to p in units of kg-m/sec, the MKS unit, we note that:

$$1 \text{ Mev}/c = \frac{1.6 \times 10^{-13} \text{ joules}}{3 \times 10^8 \text{ m/sec}}$$

$$= 0.533 \times 10^{-21} \text{ kg-m/sec}$$

since the units joule-sec/m = newton-sec = kg-m/sec, using F = ma or newtons = kg-m/sec^2.

Often, equation 1.18 is used in the form $E_T^2 = p^2 + m_0^2$ where E_T and m_0 are expressed in energy units such as Mev and p is in units of Mev/c. In other words, by solving for p in the above equation when E_T and m are expressed in Mev we actually get the quantity pc in units of Mev—but this clearly has the same numerical value as the momentum p, expressed in units of mev/c.

Example 1.20. Calculate the momentum, in units of Mev/c, of a 10 Mev proton and a 10 Mev electron, knowing that the rest mass of the proton is 938 Mev and the rest mass of the electron is 0.511 Mev.

Solution. For the proton:

$$\begin{aligned} p^2 &= E_T^2 - m_0^2 = (938 + 10)^2 - (938)^2 \\ &= (948)^2 - (938)^2 = 898704 - 879844 = 18860 \\ p &= 137.3 \text{ Mev}/c \end{aligned}$$

For the electron:

$$\begin{aligned} p^2 &= E_T^2 - m_0^2 = (0.511 + 10)^2 - (0.511)^2 \\ &= (10.511)^2 - 0.2611 = 110.48 - 0.261 = 110.22 \\ p &= 10.5 \text{ Mev}/c \end{aligned}$$

For "particles" such as the photon which have zero rest mass (see Chapter 2), the momentum is directly proportional to the energy and is given by equation 1.18 by $E_T^2 = p^2c^2$ so that $p = E_T/c$. Thus, for a photon of energy $h\nu$ we should have a momentum of $h\nu/c$. For example, a 10 Mev photon will have a momentum of 10 Mev/c.

Example 1.21. Calculate the momentum of an electron and of a photon each of which has an energy of 100 ev.

Solution. For the electron (non-relativistically):

$$E = \frac{p^2}{2m} \qquad p = \sqrt{2mE}$$

$$\begin{aligned} p &= \sqrt{2 \times 9.11 \times 10^{-31} \times 100 \times 1.6 \times 10^{-19}} \\ &= 5.40 \times 10^{-4} \text{ kg-m/sec} \end{aligned}$$

For the photon:

$$p = \frac{h\nu}{e} = \frac{100 \times 1.6 \times 10^{-19}}{3 \times 10^8} = \frac{16}{3} \times 10^{-10} = 5.33 \times 10^{-10} \text{ kg-m/sec}$$

1-8. The Mass Spectrograph. The many experiments with cathode ray tubes in the latter part of the nineteenth century led to the discovery by the German scientist Goldstein that if holes were placed in the cathode, positively charged particles emerged from these holes. These positively charged particles were found to have a charge-to-mass ratio very much smaller (by a factor of several thousand) than the value for electrons. It was soon found that these positive canal rays (after the German *Kanalstrahlen*) were positively ionized atoms of the gases present in the tube, i.e they were atoms with one or more electrons removed.

Measurements similar to those made by Thomson on electron can be made on these positive ions, and instruments utilizin these general principles are called mass spectrographs. Althoug in principle only the ratio q/m can be determined, the masse of the various atoms were known roughly from careful weighin experiments by chemists on samples containing large numbers c atoms. Since it soon also became clear, due to the work of Mill kan, that electric charges found in nature are always multiples c the charge on the electron, it was usually possible to estimate th charge on positive ions from a rough knowledge of e/m.

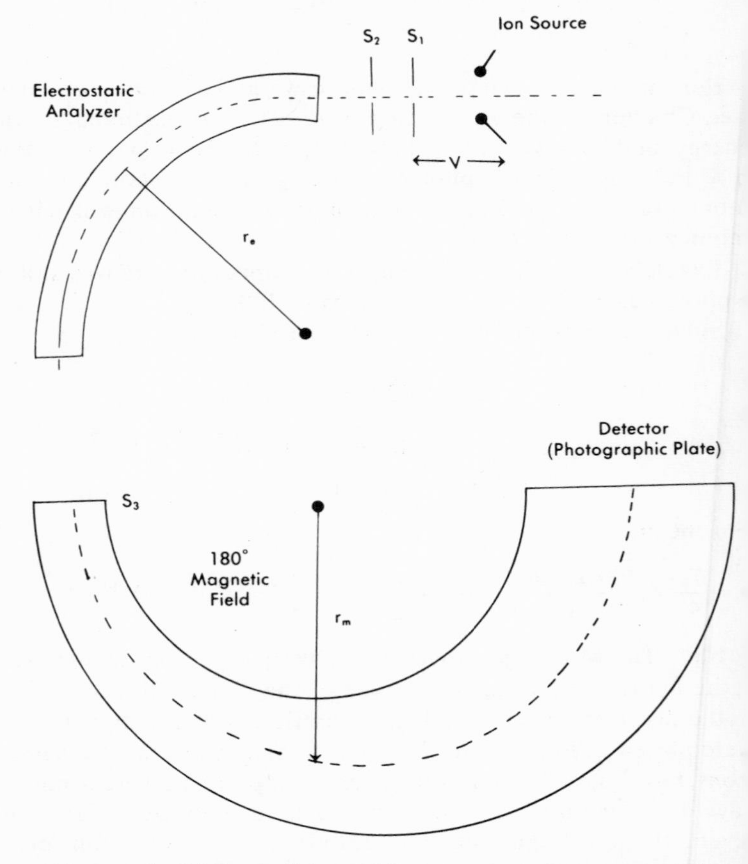

Figure 1.6. Schematic diagram of Demptster type mass spectrograph.

One type of mass spectrograph, devised by Dempster in 1935, is shown schematically in Figure 1.6. Positive ions accelerated through some potential difference V enter the system through the slits S_2 and S_1. These ions could come from a cathode ray tube, but in modern instruments other sources are usually used. Often in the instrument illustrated in Figure 1.6, positive ions are "evaporated" from a radio-frequency spark source. The positive ions then pass through parallel circular plates between which an electric field is maintained so as to bend ions of a particular charge-to-mass ratio and velocity in a circular path. This radial electrostatic analyzer acts as a velocity selector so that for a particular value of q/m, only a narrow range of velocities pass through S_3.

The ions which pass through S_3 are then further bent through 180° in a uniform magnetic field, directed into the page. They are often detected at the far end of the instrument by a photographic plate which can be removed, developed, and carefully measured. The accelerating voltage V between the source and the slits should of course be adjusted so that the velocity of the desired ions is the same as the velocity for which the electrostatic analyzer is set.

The velocity of ions of charge q and mass m accelerated through the potential V must be given by $\frac{1}{2}mv^2 = Vq$ so that:

$$v = \sqrt{\frac{2Vq}{m}} \qquad (1.19)$$

The electric force must provide the centripetal force on the ions in the electrostatic analyzer so that: $Eq = mv^2/r_e$ and:

$$v = \sqrt{\frac{Eqr_e}{m}} \qquad (1.20)$$

where E is the electric field between the plates.

In the magnetic field, the relationship:

$$Bqv = \frac{mv^2}{r_m} \qquad (1.21)$$

must hold.

Using Equations 1.20 and 1.21, it is possible to eliminate the velocity and obtain the following expression for q/m:

$$\frac{q}{m} = \frac{r_e E}{r_m^2 B^2} \qquad (1.22)$$

Example 1.22. If the accelerating voltage in a Dempster type mass spectrograph is 10,000 volts, calculate the value of the electric field required in the electrostatic analyzer for a radius r_e of 10 cm.

Solution. Using equations 1.19 and 1.20 and solving each for the velocity we get:

$$v = \sqrt{\frac{2Vq}{m}} \quad \text{and} \quad v = \sqrt{\frac{Er_e q}{m}}$$

Equating these two expressions for the velocity gives:

$$E = \frac{2V}{r_e} = \frac{2 \times 10{,}000}{0.10} = 200{,}000 \text{ volts/m}$$

Note that the value of the electric field required is independent of the value of q/m.

1-9. The Atomic Mass Unit and Nuclear Binding Energy. Although it would seem natural to measure the masses of atoms in terms of the standard kilogram, a moment's thought makes it clear that it would be much more satisfactory to measure the masses of the various atoms found in nature relative to one another using some convenient atom as the standard of atomic mass. This is done because it facilitates relative measurement and eliminates the necessity of the precise determination of all the electric and magnetic variables associated with a mass spectrometer. The results of careful measurements relating the kilogram and the atomic standard can be made and used by everyone until improved equipment and technique permit more precise measurements.

Since the nuclei of all atoms are composed of integral numbers of neutrons and protons, we might expect that if we could carefully measure the mass of each of these particles, plus that of the electron, we could calculate accurately the masses of all the atoms. This is not the case because when neutrons and protons are put together to form the various nuclei found in nature, it turns out that they weigh less in the nucleus than when they are "weighed" separately outside of the nucleus. The difference between these two masses is called the binding energy.

The binding energy per neutron or proton or the binding energy per nucleon, as it is often called, varies considerably throughout the Periodic Table as shown in Figure 1.7. The precise measurement of atomic masses amounts to the determination of the binding energy per nucleon.

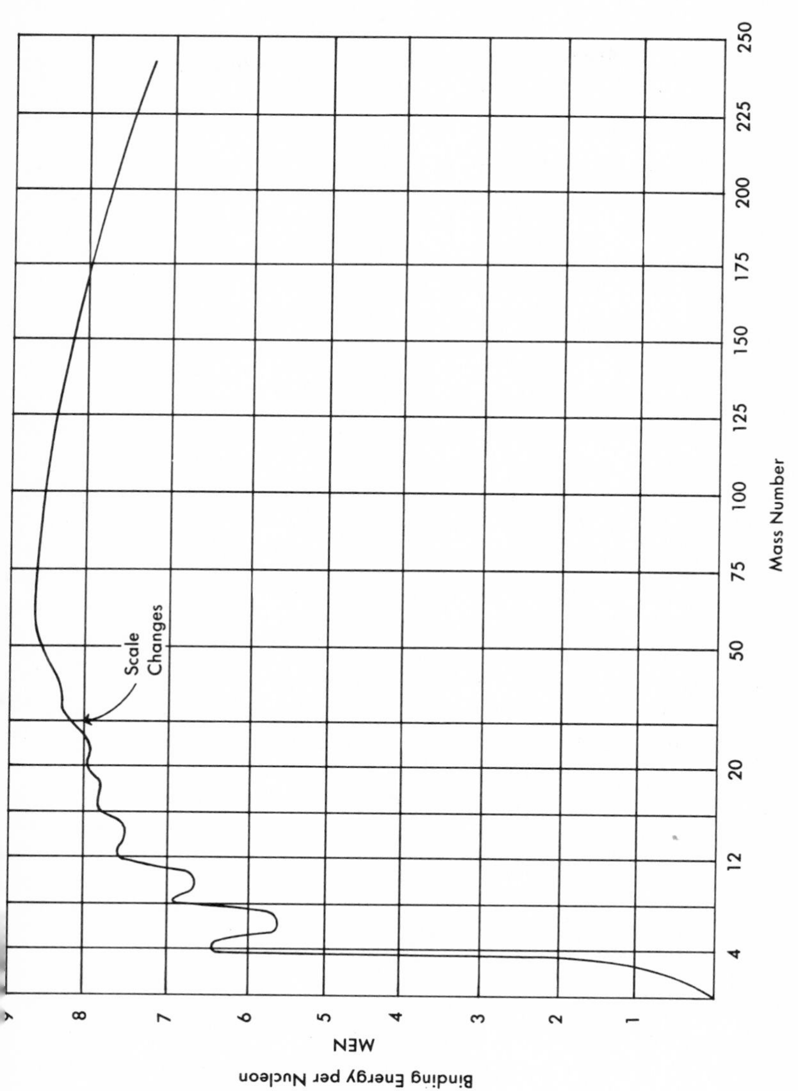

Figure 1-7. Binding Energy per Nucleon *vs.* Mass Number.

Table 1-1. Values of atomic masses from various parts of the Periodic Table.

$$^{1}H = 1.00782519 \pm 0.00000008 \text{ amu}$$
$$^{14}N = 14.00307439 \pm 0.00000017 \text{ amu}$$
$$^{16}O = 15.99491502 \pm 0.00000028 \text{ amu}$$
$$^{56}Fe = 55.9349363 \pm 0.0000043 \text{ amu}$$
$$^{235}U = 235.043915 \pm 0.000022 \text{ amu}$$

The mass of the carbon-12 atom, with six neutrons and six protons in its nucleus, has been taken as the standard of atomic masses. Its mass is defined as exactly 12.0000000. . . . atomic mass units. This choice makes the masses of most atoms come out fairly close to whole numbers, although this was not the reason it was chosen. Some elements, such as silver, occur in nature with more than one isotope. Isotopes of a particular element all have the same number of protons in their nucleus but different numbers of neutrons. Thus, silver is made up of roughly equal parts of $^{109}_{47}Ag$ and $^{107}_{47}Ag$—with the subscript being the atomic number and equal to the number of protons in the nucleus, and the superscript being the mass number which is equal to the number of neutrons plus the number of protons. Thus, ^{107}Ag has 60 neutrons while ^{109}Ag has 62 neutrons in its nucleus.

We can calculate the number of grams in one atomic mass unit using Avogadro's law, but we do this realizing that Avogadro's number (N_0) depends on the standard for atomic weights which is used. Thus, when scientists switched from a scale based on ^{16}O to the scale based on ^{12}C, Avogadro's number became slightly smaller. According to Avogadro's law, exactly 12 g of carbon should contain Avogadro's number or 6.02×10^{23} atoms, so that each carbon atom has a mass of $12 \div (6.02 \times 10^{23}) = 1.99 \times 10^{-23}$ g. One atomic mass unit should have a value, in grams, equal to one twelfth of this so that:

$$1 \text{ amu} = \frac{12}{12\,N_0} = \frac{1}{N_0} = 1.66 \times 10^{-24} \text{ g} = 1.66 \times 10^{-27} \text{ kg}$$

Atomic masses are almost always given so as to include the masses of the electrons. The mass of the electron is about 0.00055 amu, so that this is by no means a negligible amount. Some typical mass values are listed above to indicate the accuracy possible in modern mass spectroscopic measurements.

Example 1.23. If the mass of the neutron is taken as 1.008665 amu and the mass of the hydrogen atom is taken as 1.007825 amu, calculate the binding energy of the $^{56}_{26}Fe$ nucleus and the binding energy per nucleon.

Solution. Since the atomic number of iron is 26, there must be ; protons and 30 neutrons in the iron nucleus. Outside the ıcleus, 26 hydrogen atoms (including electrons) would have a ass of 26 × 1.007825 = 26.203450 amu. Similarly, the mass of) neutrons would be 30 × 1.008665 = 30.259950 amu. The ›tal mass outside the nucleus is 26.203450 + 30.259950 = ;.463400 amu. The mass of the ^{56}Fe atom is 55.934936 amu, so at the total binding energy of the nucleus is given by the difrence between these numbers. This is 0.528464 amu, which ıuals 0.528464 amu × 931.48 Mev/amu = 492.25 Mev. The nding energy per nucleon (per neutron or proton) is simply ›2.25 ÷ 56 = 8.789 Mev.

The remarkable precision of the mass measurements indicated Table 1-1 (p. 28) results primarily from a technique of measureent known as the doublet method. This method is based on the ct that it is possible to produce, in ion sources, ions of molecules mposed of hydrogen and carbon (hydrocarbon ions). These can obtained with almost any charge-to-mass ratio making it ›ssible to observe "doublets" on the photographic plate or her detection of the mass spectrometer. These doublets are two es which are very close together and have very nearly the same arge-to-mass ratio, but differ in this ratio because of a dif'ence in nuclear binding energy.

An example of such a doublet used in the determination of the ıss ^{27}Al is the following: C_2H_3–^{27}Al = 0.0419451 ± 0.0000023 ıu. A mass spectrometer permits comparison of other atoms th C^{12}, but the accuracy of such measurements is roughly)1%. The mass difference in the mass 27 doublet above is :asured to about one part in 20,000, but this represents an un·tainty of about one part in 10^7 of the mass of ^{27}Al (which is amu). Usually, the results of several such doublet measure:nts are combined to obtain masses relative to carbon-12. In s case, a series of doublet measurements are required to obtain : mass of the hydrogen atom which turns out to be 1.0078235 ıu.

Thus the mass of ^{27}Al is:

$$\text{l} = C_2H_3 - 0.0419451 = 24.000 + 3(1.0078235) - 0.0419451$$
$$= 26.981525 \text{ amu}$$

e use of the doublet technique using hydrocarbons was one of : reasons for switching in 1961 from a mass scale based on ^{16}O ng exactly 16.0 to the ^{12}C scale now in use.

Chapter 2

Wave-Particle Duality

2-1. Introduction. All of the properties of light which have been discussed in classical physics can be explained in a satisfactory way by treating light as a wave. Phenomena such as reflection, refraction, interference, diffraction, and polarization are the principal effects which we put in this category—most of these in fact require that light behave as a wave. In the following sections, we will discuss some experimental results which cannot be explained satisfactorily in terms of a wave model and which require that we think in terms of a particle or corpuscular model. This fact that light and other forms of electromagnetic radiation can exhibit wavelike properties in one situation and particle-like properties in another is often referred to as wave-particle duality. This is a fundamental property of matter on a small scale (i.e., at the atomic and molecular level), and we shall see in the latter part of this chapter that it has been found to be true also for what might be called material particles such as electrons and other atomic particles which have a non-zero rest mass.

2-2. The Photoelectric Effect. It is rather interesting and perhaps appropriate that the photoelectric effect was discovered as a by-product of the experiments done by Heinrich Hertz (1857–1894) in 1887 in which he showed that electromagnetic waves can be produced by rapidly oscillating electric charges as predicted earlier by Maxwell. The photoelectric effect is the name given to the process in which light is incident on a surface (usually metal) from which electrons are then emitted.

It might be thought that such a process could be explained in terms of a wave theory in which light is regarded as a traveling electromagnetic wave. The oscillating electric field associated with such a wave might be able to build up the energy of an electron in an atom until it finally had sufficient energy to escape. The experimental facts, however, will not support this general approach and are as follows. The number of photoelectrons is proportional to the intensity of the light, but the maximum kinetic energy of these photoelectrons is not. No matter how feeble the light source is made, the observed maximum kinetic energy of the photoelectrons is unchanged. It does depend, however, on the frequency of the light, being directly proportional to it. The time between the arrival of the light at the metal surface and the emission of a photoelectron is found to be as short as can be measured (about 10^{-9} sec) and not to depend upon the intensity. When calculations of the rate at which energy can be delivered to an atom, for a weak light source of known intensity, are made using a wave model, it is found that times of the order of 100 hours are required to build up energies of a few electron volts which typically are required to remove a photoelectron.

Something new was required to explain the photoelectric effect and it was provided in 1905 by Albert Einstein. By assuming that light could be considered as discrete bundles of energy, often called photons, with each bundle having an energy $h\nu$, where ν is the frequency and h is Planck's constant, Einstein was able to explain the experimental facts observed for the photoelectric effect. The constant h has the value 6.625×10^{-34} joule-sec and had arisen some five years earlier in the work of Max Planck (1858–1947) who found that he had to assume a similar sort of quantization in order to explain the spectrum of light emitted from a hot body. The quantity h is a fundamental constant of nature which we will use repeatedly in our discussion of systems of atomic dimensions.

Einstein's interpretation of the photoelectric effect is compactly summarized as:

$$\frac{1}{2}\mathrm{mv}^2 = h\nu_1 - h\nu_0 \qquad (2.1)$$

where $h\nu$ is the energy of the incident photon, $\frac{1}{2}\mathrm{mv}^2$ is the kinetic energy of the photoelectron, and $h\nu_0$ is the energy required to just barely remove the photoelectron, i.e. to remove it with zero kinetic energy. Thus, $h\nu_0$ gives the minimum energy of a photon capable of producing a photoelectron and ν_0 is called the threshold frequency. Often $h\nu_0$ is called the work function. Einstein's model, then, is that a photon encounters an atom and gives all of its energy to the atom. The result of this is that an electron is ejected which has a kinetic energy equal to the energy of the incident photon minus whatever energy is required to remove it from the atom and from the surface of the metal.

If this model, as represented by equation 2.1, is correct a linear relationship should exist between the maximum kinetic energy of the photoelectrons and the frequency. This is found to be the case

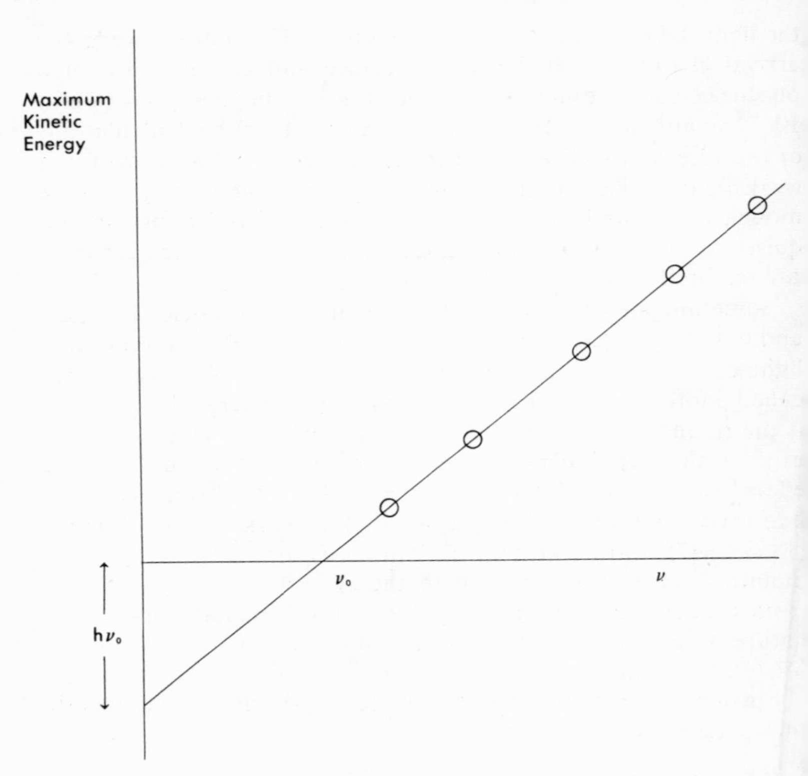

Figure 2.1. The linear relationship between photoelectron maximum kinetic energy and photon frequency.

experimentally, as is indicated in Figure 2.1. Note that the slope of this graph should be equal to Planck's constant, h, the x- or frequency-intercept should be equal to ν_0, and the y- or energy-intercept should be equal to the work function, $h\nu_0$. Such is found to be the case experimentally. The photoelectric effect can occur for visible light and for shorter-wavelength radiation, depending of course upon the magnitude of the work function. In the visible region the energy of photons is only a few electron volts, so that the work function is usually an appreciable fraction of the photon energy. For much shorter wavelengths, the photoelectron has an energy very nearly equal to the photon—in which case its energy can be used to measure the photon energy, but relativistic expressions must be used for the electron kinetic energy. In the photoelectric effect, momentum of course must be conserved, but e recoil energy is always a very small fraction of the photon ergy.

Example 2.1. Calculate the energy in electron volts of a oton in the middle of the visible region at 5000 Å, at 8000 Å in e infrared, and at 1000 Å in the ultraviolet.

Solution. At 5000 Å we get:

$$E = h\nu = \frac{hc}{\lambda}$$

$$= \frac{6.625 \times 10^{-34} \times 3 \times 10^{8}}{5 \times 10^{-7}}$$

$$= 3.975 \times 10^{-19} \text{ joules}$$

$$= \frac{3.975 \times 10^{-19} \text{ joules}}{1.6 \times 10^{-19} \text{ joules/ev}}$$

$$= 2.48 \text{ ev}$$

At 8000 Å we get:

$$E = 2.48 \times \frac{5000}{8000} = 1.55 \text{ ev}$$

ce $E \propto 1/\lambda$.

At 1000 Å:

$$E = 2.48 \times \frac{5000}{1000} = 12.4 \text{ ev}$$

Example 2.2. A particular metal surface has a work function 2.0 ev. Calculate the maximum kinetic energy of photons to be ected when light of wavelength 4500 Å is incident on the face and calculate the threshold frequency. (1 Å = 10^{-10} m)

Solution. The maximum kinetic energy is given by equation as:

$$\text{K.E.} = h\nu - h\nu_0 = \frac{hc}{\lambda} - h\nu_0$$

$$= \frac{6.625 \times 10^{-34} \times 3 \times 10^{8}}{4.5 \times 10^{-7} \times 1.6 \times 10^{-19}} - 2.0 \text{ ev}$$

$$= 2.76 - 2.0 = 0.76 \text{ ev}$$

The threshold frequency is given by:

$$h\nu_0 = 2.0 \times 1.6 \times 10^{-19}$$

$$\nu_0 = \frac{2.0 \times 1.6 \times 10^{-19}}{6.625 \times 10^{-34}} = 0.483 \times 10^{15} \text{ cycles/sec}$$

Example 2.3. Calculate the recoil kinetic energy given to an atom of mass number 50 when it absorbs a photon with an energy of 3 ev. Repeat the calculation for 10,000 ev.

Solution. The momentum of the recoiling atom must equal the momentum of the incident photon, i.e.:

$$mv = \frac{h\nu}{c} \qquad \text{so that} \qquad (mv)^2 = \frac{(h\nu)^2}{c^2}$$

Also, $h\nu = E_\gamma$, where E_γ is the photon energy. If $\frac{1}{2}mv^2 = E_R$, the recoil energy, we can write:

$$E_R = \frac{E_\gamma^2}{2mc^2}$$

Utilizing the fact that 1 amu = 931 Mev, we get for the case of a 3 ev photon:

$$E_R = \frac{3^2}{2 \times 50 \times 931 \times 10^6} = 9.7 \times 10^{-10}\ \text{ev}$$

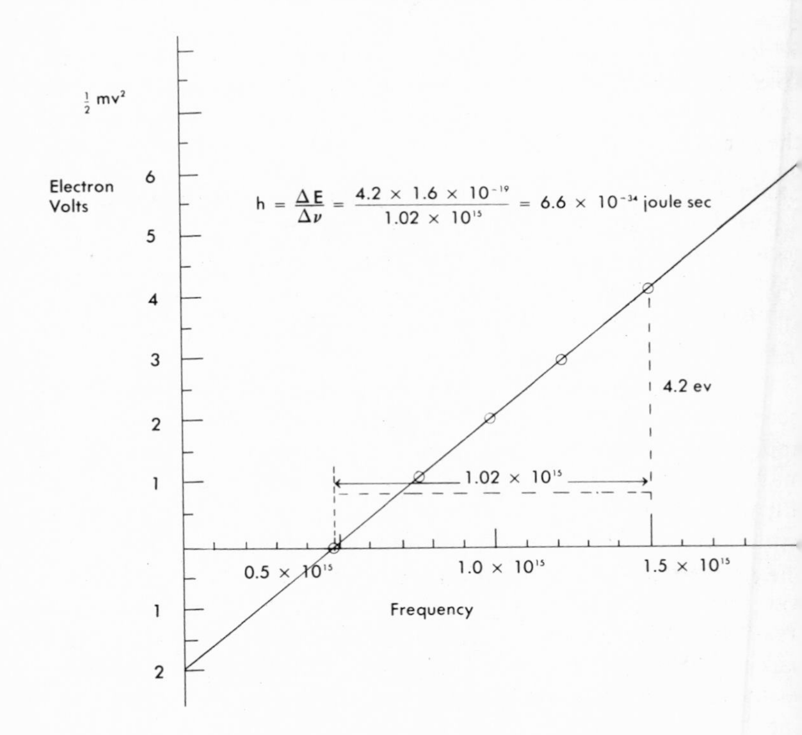

For a 10,000 ev photon we get:

$$E_R = \frac{10{,}000^2}{2 \times 50 \times 931 \times 10^6} = 1.07 \times 10^{-3} \text{ ev}$$

Note that in both cases this recoil energy is an extremely small fraction of the photon energy.

Example 2.4. The graph (p. 34) shows the maximum kinetic energy of photoelectrons released from a certain surface plotted as a function of the frequency of the incident light. From the above graph, determine the work function and Planck's constant.

Solution. The work function is given by the intercept on the energy axis and is 2 ev. Planck's constant is given by the slope of the curve which yields $h = 6.6 \times 10^{-34}$ joule-sec as indicated on the graph.

2-3. The Compton Effect. During the early 1920's, at the University of Chicago, A. H. Compton (1892–1962) discovered another process which required a corpuscular view of radiation. Compton was experimenting with very short wavelength radiation ($\lambda = 1$ Å) known as *X* rays, the production of which we will discuss in Chapter 4. It is possible to obtain a beam of *X* rays all of which have very nearly the same wavelength. In scattering these from targets of various materials, Compton observed that the scattered radiation was composed of two components—one of the same wavelength as the original *X* rays and one at a longer wavelength. The classical wave theory predicted the former but could not explain the longer wavelength. Very simply, one can imagine a wave incident on matter causing electrons to oscillate with the same frequency as the incident wave and perhaps to reradiate this same frequency in other directions, but it is not at all clear how an appreciable amount of radiation at some other frequency might be generated.

Compton also observed that the change in wavelength depended only on the angle at which the measurement was made and was independent of both the original wavelength and the material from which the radiation was scattered. Figure 2.2 illustrates the theoretical explanation which Compton used to explain this effect which bears his name.

Agreement with the basic experimental facts presented above was obtained by assuming that the incident *X* rays could be treated as particles or corpuscules of energy $h\nu$, and that these particles experienced elastic collisions with electrons in the scattering material. In an elastic collision, of course, both energy and

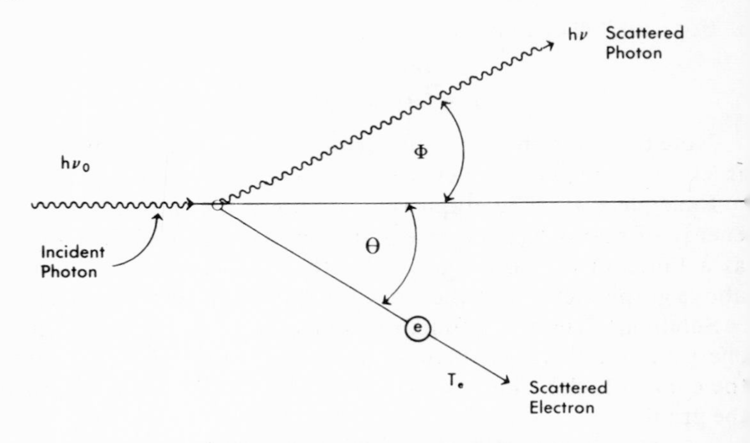

Figure 2.2. The Compton effect.

momentum are conserved, and the assumption is that the electro is essentially free and that we do not worry about it being boun to the atom. Of eourse, the electrons are bound in atoms, and th fact explains the presence of scattered radiation with wavelengt unchanged. At higher energies and shorter wavelengths, the un modified component in the scattered radiation becomes small and smaller as the assumption that the electron is unbound be comes better and better.

Applying conservation of energy to the vector diagram Figure 2.2 we get:

$$h\nu_0 = h\nu + T_e \qquad (2.$$

Conservation of momentum in the x- and y-directions respective gives:

$$\frac{h\nu_0}{c} = \frac{h\nu}{c}\cos\Phi + p\cos\Theta$$

$$\frac{h\nu}{c}\sin\Phi = p\sin\Theta \qquad (2.$$

where p is the scattered electron momentum. Squaring and addi these two equations yields:

$$\left(\frac{h\nu_0}{c} - \frac{h\nu}{c}\cos\Phi\right)^2 = p^2\cos^2\Theta$$

$$\left(\frac{h\nu}{c}\right)^2\sin^2\Phi = p^2\sin^2\Theta$$

$$\left(\frac{h\nu_0}{c}\right)^2 - \left(\frac{2h\nu_0 h\nu}{c^2}\right)\cos\Phi + \left(\frac{h\nu}{c}\right)^2\cos^2\Phi = p^2\cos^2\Theta$$

$$\left(\frac{h\nu}{c}\right)^2\sin^2\Phi = p^2\sin^2\Theta$$

Since $\sin^2\Phi + \cos^2\Phi = 1 = \sin^2\Theta + \cos^2\Theta$:

$$\left(\frac{h\nu_0}{c}\right)^2 - \left(\frac{2h\nu_0 h\nu}{c^2}\right)\cos\Phi + \left(\frac{h\nu}{c}\right)^2 = p^2 \qquad (2.4)$$

For the electron, we know that the total energy must be given by:

$$E_T^2 = p^2c^2 + (m_0c^2)^2$$

and, if T_e is the electron kinetic energy:

$$(T_e + m_0c^2)^2 = c^2p^2 + (m_0c^2)^2$$
$$T_e^2 + 2T_em_0c^2 = c^2p^2$$

or:

$$\frac{T^2}{c^2} + 2Tm_0 = p^2$$

Using equation 2.2 for T_e and equation 2.4 for p^2, we obtain:

$$\left(\frac{h\nu_0}{c} - \frac{h\nu}{c}\right)^2 + 2mc\left(\frac{h\nu_0}{c} - \frac{h\nu}{c}\right) = \left(\frac{h\nu_0}{c}\right)^2 + \left(\frac{h\nu}{c}\right)^2 - 2\left(\frac{h\nu_0}{c}\right)\left(\frac{h\nu}{c}\right)\cos\Phi$$

$$m_0c\left(\frac{h\nu_0}{c} - \frac{h\nu}{c}\right) = \frac{h\nu_0}{c}\,\frac{h\nu}{c}(1 - \cos\Phi)$$

$$\frac{1}{h\nu/c} - \frac{1}{h\nu_0/c} = \frac{1}{m_0c}(1 - \cos\Phi)$$

Since $\nu = \dfrac{c}{\lambda}$ and $\dfrac{c}{h\nu} = \dfrac{c\lambda}{hc} = \dfrac{\lambda}{h}$, we multiply through by h and get:

$$\lambda - \lambda_0 = \frac{h}{m_0c}(1 - \cos\Phi) \qquad (2.5)$$

where $\dfrac{h}{m_0c} = 0.0243 \times 10^{-10}$ m and is called the Compton wavelength.

Thus, the relatively simple assumption of an elastic collision between an incident photon and a nearly free atomic electron explains the experimental observation of a wavelength increase which is independent of initial photon energy and the nature of the scatterer. This assumption does not, however, permit any calculation of the relative probability of Compton scattering as a function of the initial photon energy and the scattered photon and electron angles. This requires a considerably more involved approach using the quantum theory of radiation which is well beyond the scope of this book.

Example 2.5. A photon of wavelength 0.1 Å is scattered through 90°. Calculate the wavelength of the scattered photon.

Solution. Using equation 2.5 and the fact that $h/\mathrm{m}_0 c$ is equal to 0.0243×10^{-10} m, we find that:

$$\lambda = \lambda_0 + \left(\frac{h}{\mathrm{m}_0 c}\right)(1 - \cos \Phi)$$

$$= 0.1 + 0.024 = 0.124 \text{ Å since } \cos \Phi = 0$$

Example 2.6. Derive an expression for the energy change experienced by a photon in the Compton process.

Solution.

$$\Delta \mathrm{E} = \frac{hc}{\lambda_0} - \frac{hc}{\lambda} = hc\left(\frac{1}{\lambda_0} - \frac{1}{\lambda}\right)$$

$$= hc\left(\frac{\lambda - \lambda_0}{\lambda_0 \lambda}\right)$$

$$= hc\left(\frac{h/\mathrm{m}_0 c\,(1 - \cos \Phi)}{\lambda_0 \lambda}\right)$$

$$= \frac{h^2}{\mathrm{m}_0 \lambda_0 \lambda}(1 - \cos \Phi)$$

The units of the expression $\dfrac{h^2}{\mathrm{m}_0 \lambda_0 \lambda}$ are $\dfrac{\text{joules}^2\text{-sec}^2}{\text{kg-m}^2}$ which can be written as $\dfrac{\text{joules}^2}{\text{kg-m}^2/\text{sec}^2} = \dfrac{\text{joules}^2}{\text{joules}}$ = joules. The expression may be written as:

$$\Delta \mathrm{E} = \frac{\mathrm{E}_0 \mathrm{E}}{\mathrm{m}_0 c^2}(1 - \cos \Phi)$$

where $\mathrm{E}_0 = \dfrac{hc}{\lambda_0}$ and $\mathrm{E} = \dfrac{hc}{\lambda}$.

Example 2.7. A 0.5 Mev photon is Compton scattered through an angle of 60°. Calculate the energy of the scattered photon and the scattered electron.

Solution. The wavelength of this photon is given by:

$\frac{hc}{\lambda_0} = 0.5 \times 10^6 \times 1.6 \times 10^{-19}$ joules $= 0.8 \times 10^{-13}$ joules. Thus,

$$\lambda_0 = \frac{hc}{0.8 \times 10^{-13}} = \frac{6.625 \times 10^{-34} \times 3 \times 10^8}{0.8 \times 10^{-13}} = 2.484 \times 10^{-12} \text{ m}$$

and $\Delta\lambda = 0.0248(1 - 0.5) = 0.0124$ Å

since $\cos \Phi = \frac{1}{2}$. The wavelength of the scattered photon is:

$$\lambda = \lambda_0 + \Delta\lambda = 0.0248 + 0.0124 = 0.0372 \text{Å}$$

The energy of the scattered photon is:

$$E_\gamma = h\nu = \frac{hc}{\lambda}$$

$$= 0.500 \text{ Mev} \times \frac{0.0248}{0.0372} = 0.333 \text{ Mev}$$

The scattered electron energy is the difference between 0.50 Mev and 0.333 Mev, or 0.167 Mev.

Example 2.8. No matter how high the energy of the incident photon, the energy of the scattered photon, at a particular angle, can never be greater than a certain amount. What is this energy, expressed in electron volts, for an angle Φ of 90°?

Solution. From equation 2.5, we see that for 90°, when $\cos \Phi = 0$, we get $\lambda - \lambda_0 = h/m_0c$. As λ_0 becomes very small for high energies, we may neglect it compared to $h/m_0c = 0.0243$, the Compton wavelength. Thus, $\lambda = 0.0243$ Å at 90° for very short wavelength photons, and the scattered photon energy is:

$$E = \frac{hc}{\lambda} \times \frac{1}{1.6 \times 10^{-19}}$$

$$= \frac{6.625 \times 10^{-34} \times 3 \times 10^8}{0.0243 \times 10^{-10} \times 1.6 \times 10^{-19}} = 0.511 \times 10^6 \text{ ev}$$

which we can also obtain somewhat more quickly by noting that $E = hc/\lambda$ and if $E = m_0c^2$, where m_0 is the rest mass of the electron:

$$\lambda = \frac{hc}{m_0 c^2} = \frac{h}{m_0 c} = 0.0243\,\text{Å}$$

Thus, E = 0.511 Mev, which is the electron rest energy.

2-4. Pair Production. With the exception of scattering wit no wavelength change, which predominates for photons in th visible region, there are three principal mechanisms by which pho tons interact with matter. We have already discussed the photo electric effect and the Compton effect. The third is called pair pro duction and provides an interesting example of the convertibilit of what we usually call matter and energy.

If a photon has energy greater than the sum of two electro rest masses (0.511 + 0.511 = 1.022 Mev), it can interact with th nucleus of an atom in such a way that it disappears and a posi tron-electron pair is produced. The positron is virtually identica to an electron, with the exception that its charge is positive, an is one of the many examples of antiparticles found in nature. Th positron produced in such an event moves off and is slowed dow by ionization and excitation in matter much as an electron is but when it is brought to rest or nearly to rest it eventually en counters an electron and the electron and positron annihilate with their rest masses being converted into two photons. Thes photons, or gamma rays as they are more often called, move o in opposite directions to conserve momentum. This annihilatio process provides another example of the conversion of matte into energy.

It is important to note that pair production cannot occur i free space and that a heavy nucleus is essential since it provide a mechanism by which energy and momentum can be conserve Since some of the gamma ray energy went into the creation of th mass of the positron-electron pair, it should be clear from equa tion 1.18 that the sum of the momenta of the electrons will b less than the photon momentum if a third object is not availabl to carry away some momentum. The mass of this nucleus is s large, however, that its kinetic energy is extremely small.

Example 2.9. A 7.0 Mev gamma ray produces a positro electron pair. If the positron and electron have the same energ calculate the energy of the electron.

Solution. If we assume that about 1 Mev is required to a count for the rest mass of the pair, this leaves about 6 Mev to b shared between positron and electron so that the electron will have a kinetic energy of 3 Mev.

Example 2.10. A 7 Mev gamma ray produces a positron-electron pair. Assume that the positron and electron come off in the same direction as the original gamma ray and show that, unless a third object takes off some momentum, it is impossible to conserve both energy and momentum in the process.

Solution. Note that this situation is the most favorable for conserving momentum without a third body. If the positron and electron were emitted at 90° to the original gamma ray direction, the nucleus would have to have the same momentum as the incident gamma ray since the positron and electron momenta are equal, opposite, and at right angles to the gamma momentum.

Assuming energy conservation, we note that each electron has a kinetic energy of 2.989 Mev if 0.511 Mev is used for the electron rest masses. Using equation 1.18 we get:

$$\begin{aligned} p^2 &= E_T^2 - m_0^2 = (3.50)^2 - 0.261 \\ &= 12.250 - 0.261 = 11.989 \\ p &= 3.473 \text{ Mev}/c \end{aligned}$$

Doubling this gives 6.946 Mev/c which is less than the gamma ray momentum of 7.0 Mev/c. If the positron and electron, for example, were given off at 60° with respect to the original gamma ray direction with the same energies, their combined momentum in the gamma ray direction would only be 3.473 Mev/c since the cosine of 60° $= \frac{1}{2}$. This would mean that the nucleus would have a momentum of $7.00 - 3.473 = 3.527$ Mev/c.

Antiparticles have been discovered for almost all the particles, such as the neutron and proton, which are known today. All annihilate violently (as do the positron and electron) when brought together with their corresponding particle although gamma rays are not always the result of this annihilation process. It is interesting to conjecture that stars in galaxies in other parts of the universe may be composed of what we call antimatter, i.e. electrons replaced with positrons and neutrons and protons in the nuclei of atoms replaced by antiprotons and antineutrons.

Figure 2.3 shows qualitatively, for the element lead, the relative probability of a photon interacting via the photoelectric effect, Compton effect, or pair production effect as a function of energy. The solid line indicates the total probability of any inter-

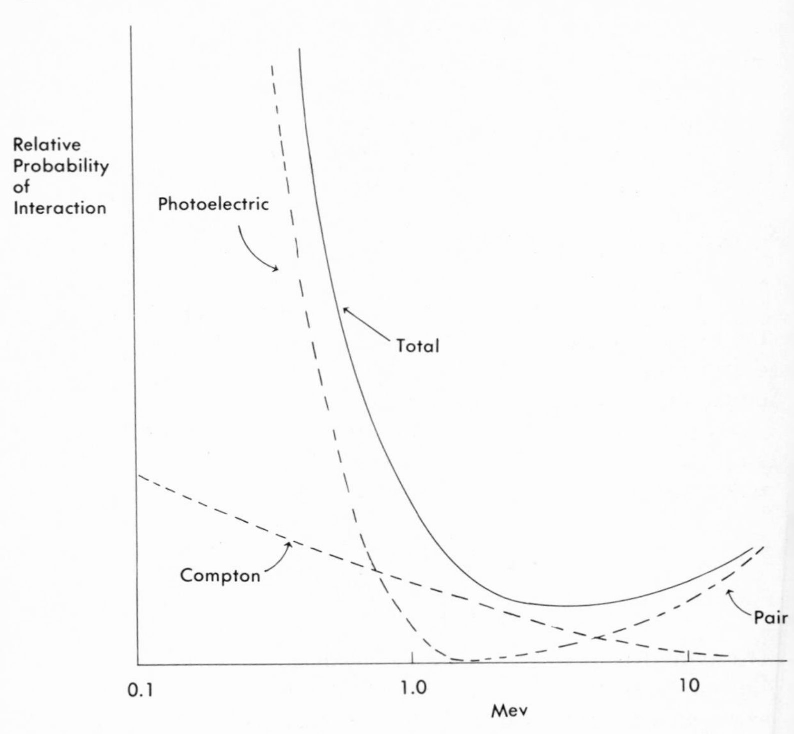

Figure 2.3. The relative probability of interaction as a function of energy for a photon in lead.

action. The photoelectric effect predominates at low energies and pair production dominates at high energies, with the Compton effect being the dominant interaction at some intermediate energy which varies from element to element.

2-5. The Wave Nature of Matter. Having found that electromagnetic radiation exhibits both wave- and particle-like properties, it is natural to ask if this situation also holds for particles with nonzero rest mass such as electrons, protons, and even whole atoms or molecules. This is indeed found to be the case, and the idea was first proposed by the Frenchman Louis de Broglie (b. 1892) in 1924. Although this duality is rather puzzling in terms of classical ideas, its extension to material particles at least provides a certain rather satisfying symmetry in nature.

De Broglie assumed that any material particle is characterized by a wavelength which is given by:

$$\lambda = \frac{h}{p} \tag{2.6}$$

where h is Planck's constant and p is the momentum of the particle. De Broglie's hypothesis provided the basis for the development of quantum mechanics which has been extremely successful in the prediction of the behavior of systems of atomic dimensions and which we will discuss briefly later in this chapter.

Before discussing how this wave nature of material particles might be detected, let us make a few calculations to see how large the wavelengths predicted by equation 2.6 might be. Let us consider an electron and a proton, each moving with a velocity of 0.1 c, where c is the velocity of light.

For the electron, we get:

$$\lambda = \frac{h}{mv} = \frac{6.625 \times 10^{-34}}{9.11 \times 10^{-31} \times 3 \times 10^{7}}$$
$$= 0.242 \times 10^{-10}\,\text{m} = 0.242\,\text{Å}$$

For the proton we get:

$$\lambda = \frac{h}{mv} = \frac{6.625 \times 10^{-34}}{1.67 \times 10^{-27} \times 3 \times 10^{7}}$$
$$= 1.32 \times 10^{-14}\,\text{m} = 0.000132\,\text{Å}$$

which is considerably smaller than the electron wavelength—in fact it is smaller by a factor of 1836; the ratio of the proton mass to the electron mass.

We see that these de Broglie wavelengths are extremely small because h is so small, but they are not hopelessly beyond measurement since the spacing of atoms in the solid state is typically of the order of a few Angstrom units.

Example 2.11. Calculate the wavelength of a golf ball having a mass of 50 g and a velocity of 20 m/sec.

Solution. Using equation 2.6 we get:

$$\lambda = \frac{h}{mv} = \frac{6.625 \times 10^{-34}}{0.05 \times 20} = 6.625 \times 10^{-34}\,\text{m}$$

This wavelength is so small that its measurement is impossible. It is about 28 orders of magnitude smaller than a typical nucleus and indicates why we see no effects of the wave nature of material particles when dealing with objects in our daily life.

Example 2.12. Show that the wavelength of an electron, expressed in Angstrom units, is given by the expression $\lambda = \sqrt{150/V}$, where V is the potential difference through which the electron is accelerated and is numerically equal to its energy in electron volts. (Assume the electron is nonrelativistic.)

Solution. Since the kinetic energy is given classically by the expression $E = p^2/2m$ so that $p = \sqrt{2mE}$ and:

$$\lambda = \frac{h}{\sqrt{2mVe}} = \sqrt{\frac{h^2}{2me} \times \frac{1}{V}}$$

then:

$$\lambda = \sqrt{\frac{(6.625 \times 10^{-34})^2}{2 \times 9.11 \times 10^{-31} \times 1.6 \times 10^{-19}} \times \frac{1}{V}}$$

$$= \sqrt{\frac{150}{V}} \times 10^{-10}\,\text{m} = \sqrt{\frac{150}{V}}\,\text{Å}$$

2-6. The Spacing of Atoms in Crystals. We have found that the de Broglie wavelength of particles such as the electron and the neutron can be about an Angstrom unit or two if the energy of the particle is in the proper range. Many substances in the solid state form crystalline materials in which the atoms are arranged in a very regular manner with very definite spacings to form a repetitive array or lattice. The regular spacing of atoms in such an array permits its use as a "natural" diffraction grating to measure the de Broglie wavelengths of such particles as electrons and neutrons.

Let us calculate the characteristic spacing between atoms in a crystal of sodium chloride (NaCl) which has a particularly simple structure. Solid NaCl, or common table salt, forms a simple cubic crystal with sodium and chlorine atoms (actually ions) located at every other corner as indicated in Figure 2.4. We shall calculate d, the spacing between adjacent sodium and chlorine atoms, using only the known density (2.163 g/cm^3) of rock salt and the atomic weights of chlorine and sodium which are 35.46 and 23.00 respectively.

If we consider a cube of rock salt with edges of 1 cm, the number of atoms along an edge is $1/d$ if d is the spacing between atoms. Thus, the total number of atoms in the 1 cm cube must be $1/d^3$.

The molecular weight of NaCl is 23.00 + 35.46 = 58.46. The number of atoms per unit volume for NaCl is given by Avogadro's law as $n = N_0 \rho / W$, where N_0 is Avogadro's number

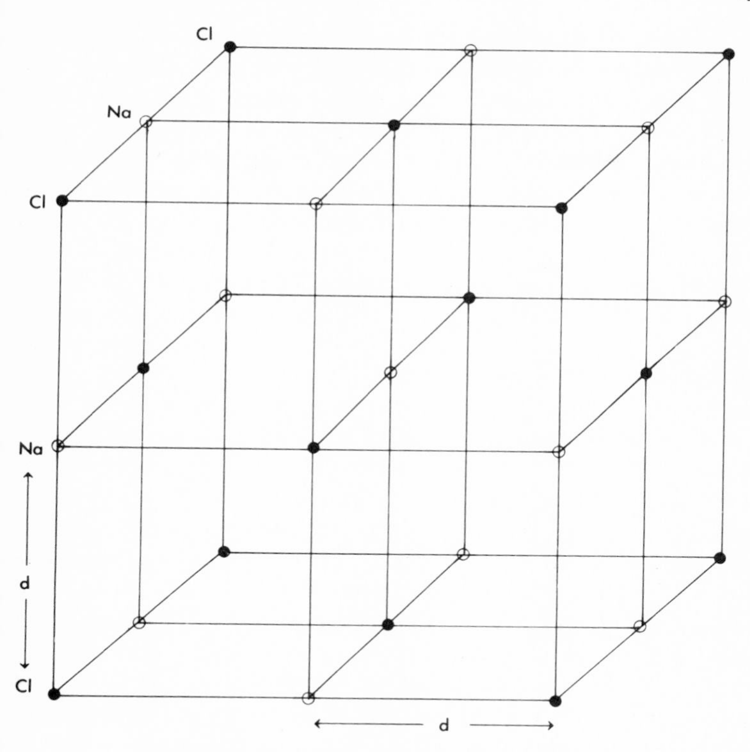

Figure 2.4. The crystal structure of NaCl.

$.02 \times 10^{-23}$ atoms/mole), ρ is the density, and W is the molecu-
r weight. Thus:

$$\frac{1}{d^3} = \frac{2 \times 6.02 \times 10^{23} \times 2.163}{58.46}$$

The factor of two in the above equation is required because we ant the number of atoms (not molecules) per unit volume, and one mole of NaCl we have Avogadro's number of Cl atoms nd Avogadro's number of Na atoms. Solving for d, we get:

$$d = 2.82 \times 10^{-8}\ \text{cm} = 2.82\ \text{Å}$$

2-7. Electron Diffraction from Natural Crystals. Experimental onfirmation of the existence of de Broglie waves was obtained the United States by Clinton J. Davisson and L. S. Germer orking at the Bell Telephone Laboratories. They announced their discovery in 1927 although as early as 1921, well before de Broglie's proposal, Davisson's experiments had shown evidence of the wave nature of electrons, which at the time went unrecognized.

Davisson and Germer observed the constructive interference of 54 ev electrons reflected or scattered from a nickel crystal. Such crystals, because of the regular arrangement of atoms and because of the close spacing of these atoms, can be used as "natural" diffraction gratings (as illustrated in Figure 2.5). Single crystals of various materials are also useful for the analysis of X rays (to be discussed in Chapter 4), which also have wavelengths of about an Angstrom unit.

In Figure 2.5, a wavefront is shown incident on a crystal. The wave interacts with many layers of atoms near the surface (although in the figure only two are used). Actually, the waves are scattered by individual atoms and the most sophisticated analysis takes account of the interference, in any desired direction, of this scattered radiation, much as though each atom in the beam acts as a small source sending out the same wavelength. It is much easier, however, to think in terms of reflections of the beam from the various layers of atoms in the crystal, as indicated in the figure. Such reflections are called Bragg reflections after the English scientist who first made this analysis in connection with his studies of X rays.

In Figure 2.5, the lower of the two rays shown must travel a distance $AB + BC$ greater than the ray reflected from the upper surface. If constructive interference is to result from waves reflected from the various planes or layers of atoms separated by a distance d, the additional distance $AB + BC$ must be equal to an integral number of wavelengths—i.e., $AB + BC$ must equal $n\lambda$. Since $AB = BC = d \sin \Theta$, the condition for constructive interference must be:

$$n\lambda = 2d \sin \Theta \tag{2.7}$$

Equation 2.7 is known as the *Bragg equation.* Although all crystals are by no means cubic, as perhaps is implied in Figures 2.4 and 2.5, many are, and it is easiest to visualize this geometry. There are many ways in which planes can be imagined in a crystal, such as the plane CD shown in Figure 2.5. Some of these will have more atoms per unit area than others, but it is possible to observe Bragg type reflections for many of these planes in most

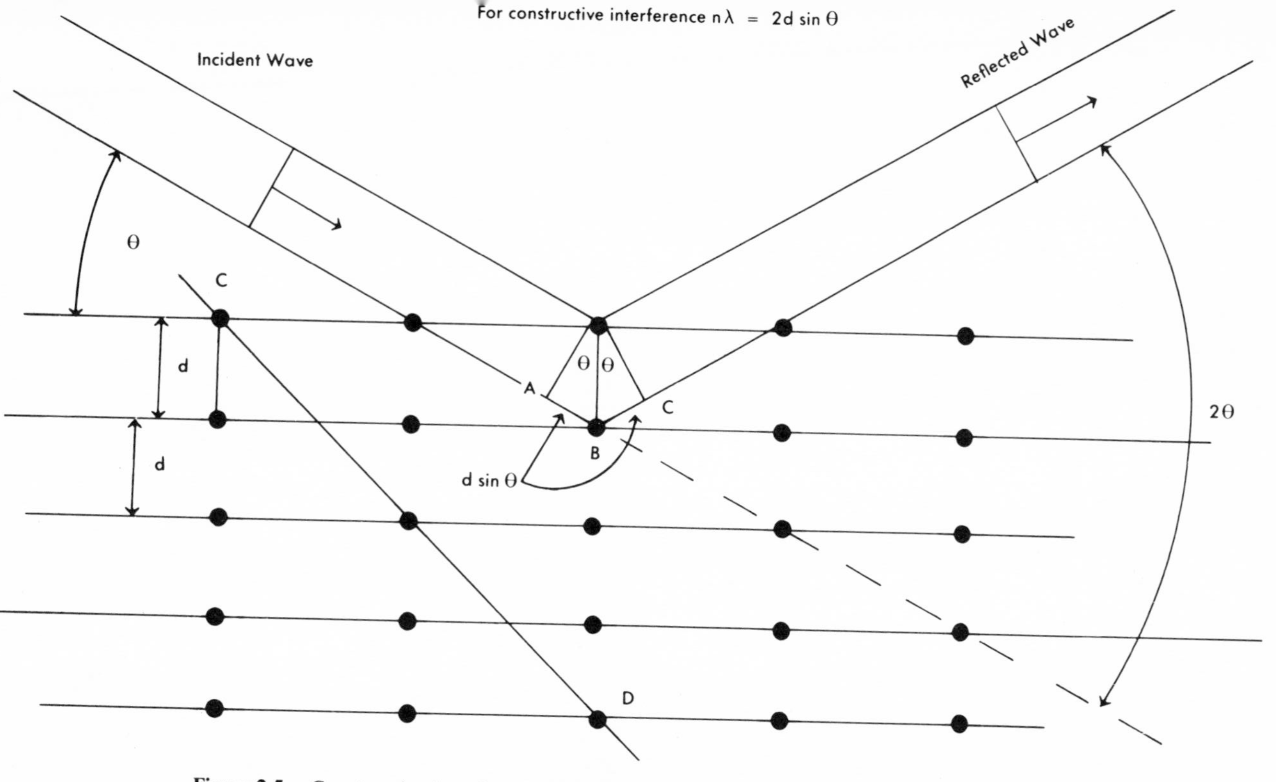

Figure 2.5. Constructive interference of de Broglie waves from layers of atoms in a crystal.

crystals. The intensity of the constructive reflection or interference will depend on the density of atoms in a particular set of planes.

Figure 2.6 illustrates another technique for utilizing the wave properties of particles, particularly for electrons. A beam of electrons is incident on a thin foil of metal which is not a single crystal, but is composed of many small crystals oriented at random. There are so many of these tiny single crystals that some of

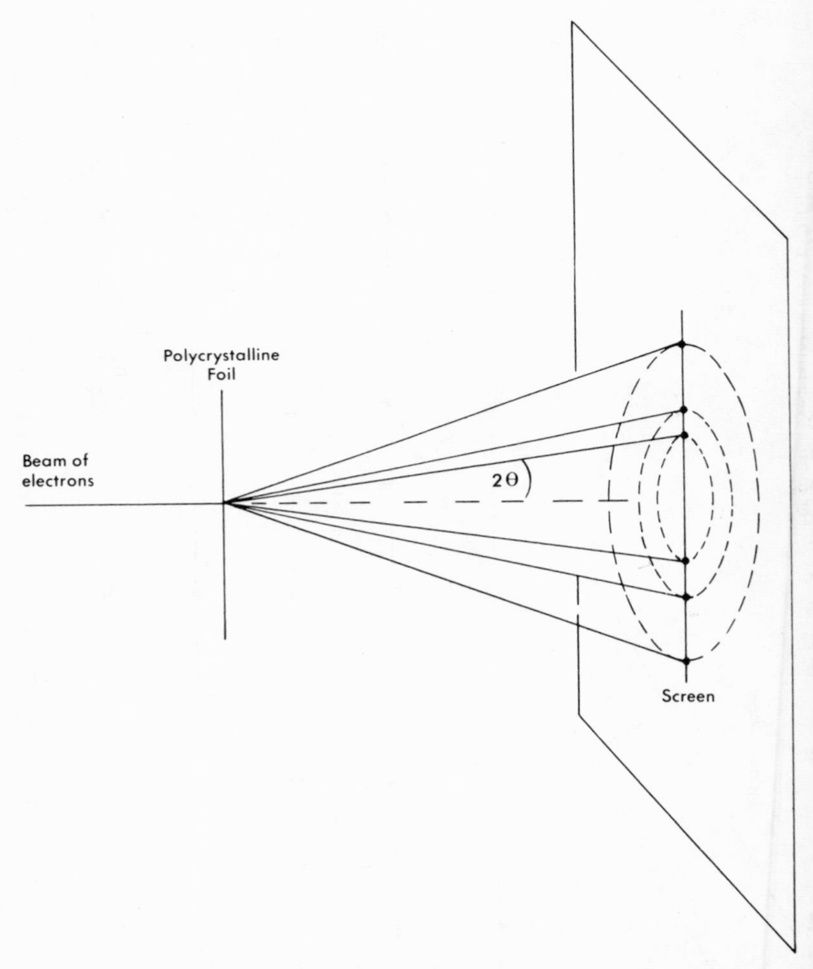

Figure 2.6. Electron diffraction from a thin polycrystalline metal foil.

ıem are oriented in just the right manner to produce Bragg re-
ection for a particular set of planes. This results in a cone of
ectrons leaving the foil, having been reflected at the Bragg angle.
hese electrons together form a circle on a fluorescent screen or
ıotographic film placed at right angles to the beam. Thus, each
rcle appearing on the screen corresponds to a different set of
anes in the crystals.

Example 2.13. Calculate the de Broglie wavelength of the
ev electrons used by Davisson and Germer.

Solution. We have shown that $\lambda = \sqrt{150/V}$ Å. Therefore:

$$\lambda = \sqrt{150/54} = 1.67 \text{ Å}$$

Example 2.14. Calculate the angle at which Davisson and
ermer must have observed the Bragg reflection of 54 ev elec-
ons (λ = 1.67 Å) from a nickel crystal with a spacing d of
15 Å. (Use n = 1.)

Solution.

$$\sin \Theta = \frac{\lambda}{2d} = \frac{1.67}{2 \times 2.15} = 0.3884$$

$$\Theta = 23^\circ$$

the Bragg equation, 23° indicates that the angle with respect to
original electron beam must be 2Θ or 46°.

Example 2.15. The energies of slow neutrons are often
asured using Bragg reflection and crystals of various types.
-called "thermal" neutrons are in thermal equilibrium with
ir surroundings and are just as likely to gain or lose energy
collisions with their surroundings. Such neutrons have a most
obable kinetic energy of about 1/40 ev. Calculate the wave-
gth of 1/40 ev neutrons.

Solution.

$$\lambda = h/p = h/\sqrt{2mE}$$

ce $E = p^2/2m$:

$$\lambda = \frac{6.625 \times 10^{-34}}{\sqrt{2 \times 1.67 \times 10^{-27} \times 1/40 \times 1.6 \times 10^{-19}}}$$

$$= 1.81 \times 10^{-10} \text{ m} = 1.81 \text{ Å}$$

-8. The Uncertainty Principle. An interesting and important
ult of the de Broglie hypothesis and the wave-particle duality
hat there must exist a theoretical limit to the precision with

which the position and momentum of a particle can be determined simultaneously. If a particle is to be described in terms of its wavelength, this wavelength really specifies the size of the particle. It is unreasonable to attempt to give the particle's position with an uncertainty too much less than one wavelength because, if we are to define a wavelength in any meaningful way, the wave must extend in space over at least a few wavelengths.

Let us estimate in a very crude way what the relationship must be between the uncertainty in the position and the uncertainty in the momentum. To do this, let us consider the experiment shown in Figure 2.7. A beam of electrons, which we represent as plane waves, are incident on a single slit whose width, Δx is a bit larger than the de Broglie wavelength of the electrons. The angle Θ at which the first minimum is observed in the diffraction pattern on the screen at the right is given by $\sin\Theta = \dfrac{\lambda}{\Delta x}$. It will be recalled

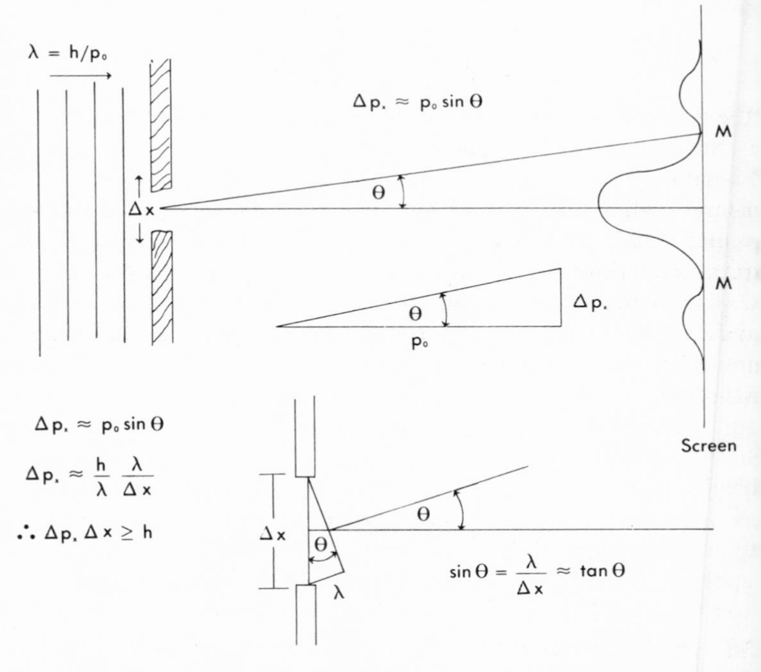

Figure 2.7. Diffraction of de Broglie waves by a single slit illustrating the Heisenberg uncertainty principle.

that if this is the case, the "waves" from the upper half of the slit will cancel those from the lower half at the points M in Figure 2.7 since, for each point in the upper half of the slit, there will be a point in the lower half of the slit for which the waves arriving at the points M will be one-half wavelength out of phase.

This approach utilizing the wave nature of the electrons does not permit us to say where any one electron will hit the screen. All we can say is that if many electrons go through the slit, their distribution on the screen will be as indicated in Figure 2.7 with a central maximum and successive minima and maxima on either side (just as is the case with light incident on a single slit whose width is of the order of the wavelength of the light). Thus, no matter how carefully we do the experiment, we run into a fundamental limitation on the accuracy with which we can specify p_x, the component of the electrons' momentum perpendicular to p_0.

If Θ is small, this uncertainty in p_x is given by $\Delta p_x = p_0 \sin \Theta$. The condition for destructive interference at M is that $\sin \Theta = \lambda / \Delta x$ where Δx is the slit width. Thus:

$$\Delta p_x \approx p_0 \sin \Theta \approx \frac{h}{\lambda} \frac{\lambda}{\Delta x} \approx \frac{h}{\Delta x}$$

where we have used the fact that $\lambda = h/p$. The relationship:

$$\Delta p \Delta x \geq h \tag{2.8}$$

is known as the *Heisenberg uncertainty principle* after the German physicist Werner Heisenberg (b. 1901), who first called attention to this relationship in 1927.

We have already pointed out that it doesn't make much sense to attempt to measure the position of a particle to an accuracy much greater than the de Broglie wavelength. Such an attempt would result in a large uncertainty in the wavelength and a resulting uncertainty in the momentum which is given by $p = h/\lambda$. Similarly, if we specify the momentum very accurately, we need to specify the wavelength very accurately and this implies an uncertainty in x which must be at least several wavelengths since an accurate wavelength determination implies measurements extending over a region of spacing containing many wavelengths. It should be emphasized that the uncertainty relationship is a fundamental property of nature and thus has nothing to do with the limitations which may exist for any particular set of instruments used for making measurements.

Example 2.16. Use the uncertainty principle to estimate th momentum, and thus the energy, of an electron confined to region having dimensions of 10^{-8} cm (such as an atom).

Solution. Using $\Delta x = 10^{-10}$ m we get:

$$\Delta p = \frac{h}{\Delta x} = \frac{6.625 \times 10^{-34}}{10^{-10}} = 6.625 \times 10^{-34} \text{ kg-m/sec}$$

If we assume that the momentum can be no smaller than the u certainty in the momentum and set $\Delta p \approx p$, we get for the u certainty in the energy (using $E = p^2/2m$):

$$\Delta E = \frac{p^2}{2m} = \frac{6.625 \times 10^{-24}}{2 \times 9.1 \times 10^{-31}}$$
$$= 2.41 \times 10^{-17} \text{ joules}$$
$$= \frac{2.41 \times 10^{-17} \text{ joules}}{1.6 \times 10^{-19} \text{ joules/ev}}$$
$$= 150 \text{ ev}$$

This indicates that considerable uncertainty exists in any attem to specify the momentum or position or energy of an electr confined to a system of atomic dimensions. It should be em phasized that such calculations involving the uncertainty princip are to be considered quite rough and only good to within perha an order of magnitude.

It is interesting, however, to make a similar calculation for electron confined in a volume of dimensions 10^{-15} m, which roughly a nuclear radius. Such an argument can be used to sh that electrons cannot exist, as such, in the nucleus and therefo must be created in the process of beta decay, in which electro are emitted by nuclei.

Example 2.17. Estimate the uncertainty in the energy of electron and a proton confined to a nucleus of radius 4×10^{-15}

Solution. We use $\Delta x = 4 \times 10^{-15}$ m and get for the u certainty in the momentum:

$$\Delta p = \frac{h}{\Delta x} = \frac{6.625 \times 10^{-34}}{4 \times 10^{-15}}$$
$$= 1.66 \times 10^{-19} \text{ kg-m/sec}$$
$$= \frac{1.66 \times 10^{-19}}{0.533 \times 10^{-21}} \text{ Mev}/c$$

since $\quad 1 \text{ Mev}/c = 0.533 \times 10^{-21}$ kg-m/sec:

$$\Delta p \approx p = 311 \text{ Mev}/c$$

To get the uncertainty in the energy, we use the relativistic relationship $E_T^2 = p^2c^2 + (m_0c^2)^2$ noting that $(m_0c^2)^2$ is about 0.261 and negligible compared to $(311)^2$. Thus $\Delta E_T = 311$ Mev, and the kinetic energy is about one-half an Mev less, since the total energy is equal to the kinetic energy plus the rest energy (which is 0.511 Mev for the electron). Again, if we assume that the energy must be at least as great as the uncertainty in the energy, the uncertainty principle tells us that to localize an electron in the nucleus means that it must have an energy of some 300 Mev.

Since the binding energy of the neutrons and protons in the nucleus averages about seven or eight Mev, it would appear impossible for electrons to exist in the nucleus for any length of time. It should be noted that a more rigorous derivation of the uncertainty principle gives:

$$\Delta p \Delta x \approx \frac{\hbar}{2\pi}$$

where $\hbar = h/2\pi$.

For the proton in the same nucleus we of course get the same Δp which is about 311 Mev/c. When we estimate the uncertainty in the energy using $E_T^2 = p^2c^2 + (m_0c^2)^2$, however, we find that for the proton we cannot neglect $(m_0c^2)^2$ since the rest mass of the proton is 938 Mev. Thus:

$$E_T^2 = 311^2 + 938^2 = 96{,}721 + 879{,}844$$

$$E_T = 988 \text{ Mev}$$

and the kinetic energy is:

$$\text{K.E.} = 988 - 938 = 50 \text{ Mev}$$

This value, although much smaller than 311 Mev, is still large, but if the uncertainty principle is used in the form $\Delta p \Delta x \approx \hbar/2$ this result becomes extremely small. This is because p goes from 311 Mev/c to 24.7 Mev/c, and when this is combined in quadrature with the rest mass of the proton, the result is about 0.3 Mev. For the electron case, the more rigorous form of the uncertainty principle gives simply 24.7 Mev for the uncertainty in the energy in the above example and this is still large for the nucleus.

Example 2.18. Calculate the uncertainty in the momentum of a golf ball of mass 50 grams and velocity 10 m/sec if we wish to specify its position to 0.001 mm, i.e. to 10^{-6} m.

Solution.

$$\Delta p = \frac{h}{\Delta x} = \frac{6.625 \times 10^{-34}}{10^{-6}} = 6.625 \times 10^{-28}$$

When we compare this with the momentum by taking the ratio:

$$\frac{\Delta p}{p} = \frac{6.625 \times 10^{-28}}{0.05 \times 10} = 13.2 \times 10^{-28}$$

we see that as far as the objects with which we deal in our daily life are concerned, the uncertainty principle has absolutely no effect. Only for particles such as electrons with very small masses and correspondingly small momenta must we worry about the uncertainty principle.

The uncertainty principle can also be expressed in terms of the uncertainty in the energy of a particle and the time at which we specify this energy. If we start with $\Delta p \Delta x \approx h$ and replace Δx by $v\Delta t$ and Δp by $m\Delta v$, we get $mv\Delta v\Delta t \approx h$. Since $E = \frac{1}{2}mv^2$, we can differentiate E to get $\Delta E = mv\Delta v$ so that $\Delta v = \Delta E/mv$ giving:

$$\Delta E\Delta t \approx h \tag{2.9}$$

We should get results comparable to those in the examples above if we use equation 2.9 and calculate ΔE by using a Δt obtained by estimating the time a particle such as an electron might take to move across a nucleus.

Example 2.19. Use equation 2.9 to estimate the uncertainty in the energy of an electron confined to a nucleus of radius 4×10^{-15} m.

Solution. If we use Δt as the time required for an electron to move a distance equal to the nuclear radius and if we assume that the velocity of the electron is only slightly less than the velocity of light we get:

$$\Delta t = \frac{4 \times 10^{-15}}{3 \times 10^{8}} = 1.33 \times 10^{-23} \text{ sec}$$

and:

$$\Delta E = \frac{h}{\Delta t} = \frac{6.625 \times 10^{-34}}{1.33 \times 10^{-23} \times 1.6 \times 10^{-13}}$$

When we use the fact that 1 Mev $= 1.6 \times 10^{-13}$ joules:

$$\Delta E \approx 300 \text{ Mev}$$

Chapter 3

Atomic Spectra

3-1. Introduction. Soon after Rutherford's discovery of the nuclear atom, the first successful atomic model was proposed by the young Danish scientist Niels Bohr (1885–1962), who had come to work in Rutherford's laboratory. This model was applicable only to simple two-body systems such as the hydrogen atom and to such systems as the singly ionized helium atom, the doubly ionized lithium atom, etc. This model, in its most simple form, considered the hydrogen atom, for example, as a stationary positively charged proton about which revolved a single negatively charged electron—the electrical attraction between the two particles providing the force which holds the atom together. With the exception of certain assumptions introduced by Bohr to provide agreement with the experimentally observed line spectrum of discrete frequencies for hydrogen, this was strictly a classical model. Despite its simplicity, it provided remarkable agreement with experiment in almost all respects.

3-2. The Bohr Model. Bohr's model of the hydrogen atom can be presented in terms of four assumptions or postulates which follow.

i. The electron moves in a circular path about the nucleus and is held in this path by the electrical force of attraction between the electron and the positively charged nucleus. If the charge on the nucleus is Ze, we express the above statement mathematically as:

$$\frac{mv^2}{r} = \frac{kZe^2}{r^2} \tag{3.1}$$

in which the electric force is equated to the centripetal force required to keep the electron in its circular path. For simplicity, we replace $1/4\pi\epsilon_0$ by k.

ii. Not all orbits or circular paths are allowed, but only those for which the angular momentum of the electron is given by:

$$mvr = \frac{nh}{2\pi} \tag{3.2}$$

where n is an integer. Thus, only those orbits for which the angular momentum is an integral multiple of $h/2\pi$ are allowed. The constant h, which has the value 6.62×10^{-34} joule-sec, is a fundamental constant of nature which we have already discussed in Chapter 2. We shall have more to say about it in Chapter 4.

iii. The electron, as it moves about the nucleus, is accelerated toward the center of its circular orbit. Classical physics predicts that accelerated charges should radiate, i.e., give off light. If this were the case, the electron should lose energy and gradually spiral in toward the nucleus, resulting in the effective collapse of the atom. This doesn't happen, so Bohr postulated that the allowed orbits, as specified under postulate *ii*, are also nonradiating orbits in which the energy of the system does not change.

iv. Electromagnetic radiation is emitted by the atom only when the electron suddenly changes or jumps from one orbit to another of lower energy. If E_i is the total energy of the initial orbit and E_f is the total energy of the final orbit, the frequency of the radiation emitted in the transition is given by:

$$h\nu = (E_i - E_f) \tag{3.3}$$

Using these postulates it is possible to make some simple calculations which should give the frequencies of the light emitted by the hydrogen atom. We cannot, however, predict anything about the relative probability of the transitions from one allowed orbit to another, and hence nothing about the intensities of the various frequencies or "lines" in the spectrum of hydrogen.

Using equations 3.1 and 3.2, we can eliminate the velocity v from equation 3.2:

$$v^2 = \frac{n^2h^2}{(2\pi)^2m^2r^2}$$

and from equation 3.1

$$v^2 = \frac{Zke^2}{mr}$$

Equating these two expressions for v^2 gives:

$$r = \frac{n^2h^2}{4\pi^2Zkme^2} \tag{3.4}$$

Setting $n = 1$ in equation 3.4 gives $r = 0.529 \times 10^{-10}$ m for what is called the first Bohr orbit.

To use equation 3.3, we must obtain an expression for the total energy of the electron-nucleus system. Equation 3.1 leads directly to the relationship $\frac{1}{2}mv^2 = Zke^2/2r$ which is the kinetic energy. Using Coulomb's law, the potential energy of the system is P.E. = $-Zke^2/r$ which is negative because the zero for potential energy is taken at $r = \infty$. The total energy of the system is the sum of the kinetic and potential energies and is:

$$E = \text{K.E.} + \text{P.E.} = \frac{-Zke^2}{2r} \qquad (3.5)$$

When the expression for the radius, given in equation 3.4, is substituted in the above expression for the total energy we get:

$$E = \frac{-2\pi^2 k^2 me^4 Z^2}{n^2 h^2} \qquad (3.6)$$

This gives us the total energy in terms of known constants and permits us to use equation 3.3 to obtain an expression for the frequencies or wavelengths which this theory predicts that the hydrogen atom should emit. We obtain:

$$\nu = \frac{E_i - E_f}{h} = \frac{2\pi^2 k^2 me^4 Z^2}{h^3}\left(\frac{1}{n_f^2} - \frac{1}{n_i^2}\right) \qquad (3.7)$$

Equation 3.7 gives good agreement with experimental measurements of the light emitted by the hydrogen atom and other "hydrogen-like" atoms.

The integer n is called a quantum number and it should be clear from equation 3.6 that the effect of quantizing or limiting the angular momentum values has also had a similar effect on the energy. This situation is represented in Figure 3.1 which is an energy level diagram for the hydrogen atom.

Equation 3.6 indicates that the allowed energy levels in the hydrogen atom vary as $1/n^2$. The energy required to ionize the hydrogen atom, i.e., to remove the electron completely from the atom, is about 13.6 ev and is equal to the energy difference between the levels corresponding to $n = 1$ and $n = \infty$ (as shown in Figure 3.1).

Figure 3.2 shows many of the possible transitions that can occur between energy levels in the hydrogen atom organized according to the level at which the transition ends, n_f. For each

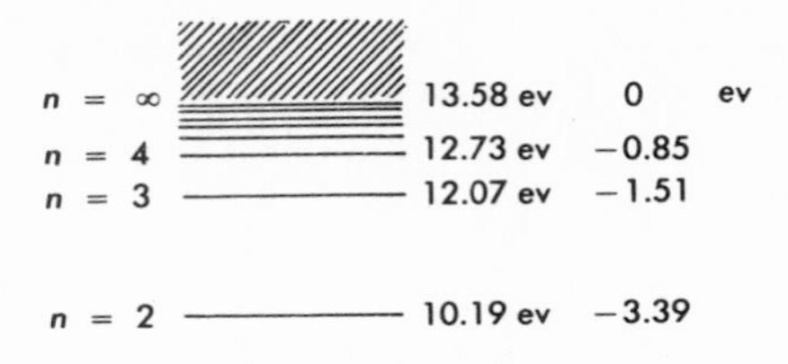

Figure 3.1. Energy level diagram for the hydrogen atom.

series, all transitions have the same n_f. The Lyman series, fo which $n_f = 1$ and for which the energy differences are greatest, i composed of transitions producing light in the ultraviolet regio of the electromagnetic spectrum. The Balmer series ($n_f = 2$) is i the visible region, while the others produce light in the infrare portion of the spectrum.

It is interesting to make a rather oversimplified calculatio based on the wave nature of the electron in which we assume tha the allowed Bohr orbits are those which permit standing waves t be set up at that radius by an electron of wavelength λ, i.e., th wavelength fits into the circumference of the circular orbit a integral number of times. The condition for this is that $2\pi r = n\lambda$ Since $\lambda = h/mv$, we get:

$$2\pi r = \frac{nh}{mv} \quad \text{or} \quad mvr = \frac{nh}{2\pi}$$

which was one of the postulates used by Bohr.

Example 3.1. Evaluate the quantity $\frac{2\pi^2 k^2 me^4 Z^2}{h^2}$ from equatio 3.6 in units of electron volts. Use Z = 1.

Solution. Using $k = 8.99 \times 10^9$, $m = 9.108 \times 10^{-31}$, e

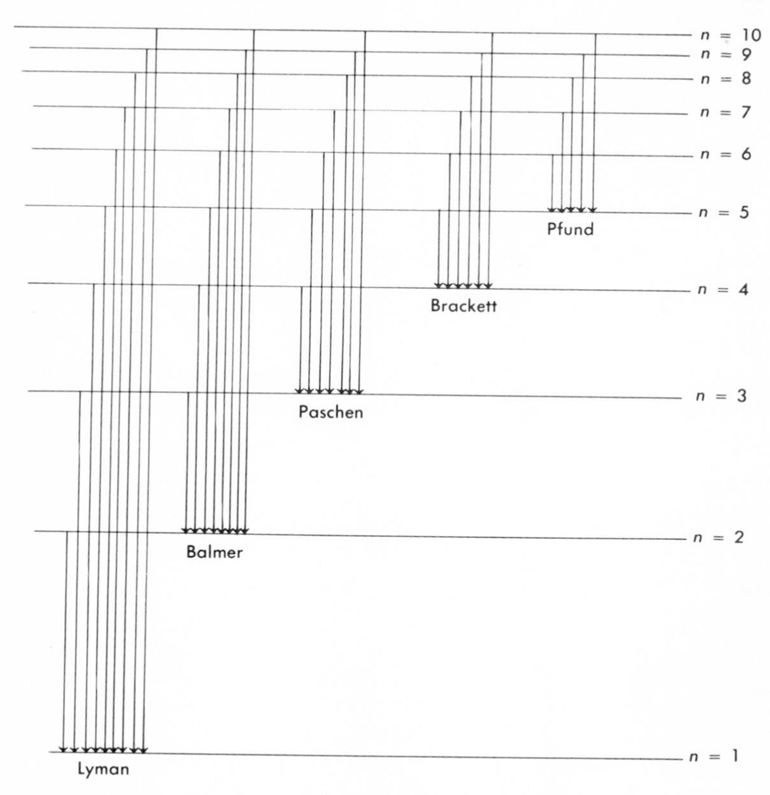

Figure 3.2. Transitions leading to various "series" of the hydrogen spectrum (levels not to scale).

1.602×10^{-19}, and $h = 6.625 \times 10^{-34}$ we get:

$$E = \frac{2 \times (8.99)^2 \times 10^{18} \times (3.14)^2 \times 9.108 \times 10^{-31} \times (1.602)^4 \times 10^{-76}}{(6.626)^2 \times 10^{-68} \times 1.602 \times 10^{-19}}$$

where the last term in the denominator converts from joules to electron volts, and:

$$E = 13.60 \text{ ev}$$

Equation 3.7 indicates that this is the energy required to remove an electron from the lowest state in the hydrogen atom, corresponding to $n = 1$, i.e., the ionization potential.

Example 3.2. Using the results of the problem above, calculate the energy difference between the levels in the hydrogen corresponding to $n = 6, 5, 4, 3$, and 2 and the lowest, or ground state, level corresponding to $n = 1$.

Solution. Using equation 3.7, we get $\Delta E = 13.6 - 13.6/n^2$ where $13.6/n^2$ gives the energy required to remove the electron from the atom when it is in the level corresponding to n.

For n values of 6, 5, 4, 3, and 2 we get, respectively:

$$n = 6 \qquad E = 13.6 - 13.6/36 = 13.6 - 0.378 = 13.22 \text{ ev}$$
$$n = 5 \qquad E = 13.6 - 13.6/25 = 13.6 - 0.544 = 13.056 \text{ ev}$$
$$n = 4 \qquad E = 13.6 - 13.6/16 = 13.6 - 0.850 = 12.750 \text{ ev}$$
$$n = 3 \qquad E = 13.6 - 13.6/9 = 13.6 - 1.511 = 12.089 \text{ ev}$$
$$n = 2 \qquad E = 13.6 - 13.6/4 = 13.6 - 3.40 = 10.20 \text{ ev}$$

This relative spacing of the levels in hydrogen is indicated in Figure 3.1.

Example 3.3. Obtain an expression for $1/\lambda$ similar to equation 3.7, evaluating the expression outside the parentheses so that $1/\lambda$ is expressed in Å^{-1}.

Solution. Since we have $\Delta E = 13.6(1/n_f^2 - 1/n_i^2)$ and $E = h\nu$ we get:

$$\nu = \left(\frac{13.6 \times 1.6 \times 10^{-19}}{6.625 \times 10^{-34}}\right)\left(\frac{1}{n_f^2} - \frac{1}{n_i^2}\right) \text{sec}^{-1}$$

$$= (3.29 \times 10^{15})\left(\frac{1}{n_f^2} - \frac{1}{n_i^2}\right) \text{sec}^{-1}$$

from which:

$$\frac{1}{\lambda} = \left(\frac{3.29 \times 10^{15}}{3 \times 10^8}\right)\left(\frac{1}{n_f^2} - \frac{1}{n_i^2}\right)$$

$$= (1.097 \times 10^7)\left(\frac{1}{n_f^2} - \frac{1}{n_i^2}\right) \text{m}^{-1}$$

$$= (1.097 \times 10^{-3})\left(\frac{1}{n_f^2} - \frac{1}{n_i^2}\right) \text{Å}^{-1}$$

using the relationship $1/\lambda = \nu/c$.

The number 1.097×10^{-3} Å^{-1} is often called the *Rydberg constant*.

Example 3.4. Calculate the wavelength in Angstrom units of the first two transitions in the series ending at the level for which $n = 6$, i.e., $n = 7$ to 6 and 8 to 6 for quadruply-ionized boron ($Z = 5$).

Solution.

$$\frac{1}{\lambda} = Z^2 \times 1.097 \times 10^{-3}\left(\frac{1}{n_f^2} - \frac{1}{n_i^2}\right)$$

For the transition $n = 7$ to $n = 6$ we get:

$$\frac{1}{\lambda} = (Z^2 \times 1.097 \times 10^{-3})\left(\frac{1}{36} - \frac{1}{49}\right)$$

$$= (25 \times 1.097 \times 10^{-3})\,(0.0278 - 0.0204)$$

$$= 0.0002021\ \text{Å}^{-1}$$

$$\lambda = 4948\ \text{Å}$$

For the transition $n = 8$ to $n = 6$:

$$\frac{1}{\lambda} = (25 \times 1.097 \times 10^{-3})\left(\frac{1}{36} - \frac{1}{64}\right) = 0.0003331$$

$$\lambda = 3002\ \text{Å}$$

3-3. Motion of the Nucleus. A small correction can be made o the semiclassical Bohr model of the atom which provides even etter agreement with experiment. This correction takes account f the fact that, in such a model, the electron and proton must otate about their common center of mass. We have assumed that ne electron revolves about a stationary proton, which would be ne case if the mass of the proton were infinite. It can be shown nat to account for this motion of the nucleus, one must replace ne mass of the electron in equations 3.6 and 3.7 with a quantity nown as the reduced mass. This is given by the relationship:

$$\frac{1}{\mu} = \frac{1}{m} + \frac{1}{M} \tag{3.8}$$

where m and M are the masses of the electron and nucleus respectively.

Example 3.5. Calculate the reduced mass for the electron in ne hydrogen atom and calculate the change in wavelength in the rst line of the Balmer series ($n = 3$ to $n = 2$) produced by this ffect. Assume that the wavelength corresponding to this transion is 6560 Å with no correction.

Solution. If we take the mass of the electron as one and the mass of the proton as 1836 times the mass of the electron, we get:

$$\frac{1}{\mu} = \frac{1}{1} + \frac{1}{1836} = 1.0 + 0.000545 = 1.000545$$

$$\mu = 0.999456 \text{ m}$$

where m is the electron mass, which shows that this is not a large effect for hydrogen. Since the energy difference between levels depends directly on m, we can calculate the change in wavelength for the first line in the Balmer series by dividing 6560 Å by 0.999456 since a reduction in mass reduces the energy and thus increases the wavelength. Thus:

$$\Delta\lambda = \frac{6560}{0.999456} - 6560 = 6563.57 - 6560$$

$$= 3.6 \text{ Å}$$

which is easily measured with moderately good spectroscopic equipment.

Example 3.6. Calculate the wavelength change to be expected for the first line of the Balmer series for deuterium as compared with hydrogen. Deuterium is the isotope of hydrogen of mass number two and is identical with ordinary hydrogen except that its nucleus is twice as heavy (being composed of a neutron and a proton). Assume that the wavelength of this line is 6560 Å as calculated according to the Bohr theory, not taking into account the motion of the nucleus.

Solution. The reduced mass for deuterium is:

$$\frac{1}{\mu} = \frac{1}{1} + \frac{1}{2 \times 1836} = 1.0002723$$

$$\mu = 0.9997277$$

The reduced mass for ordinary hydrogen is:

$$\frac{1}{\mu} = \frac{1}{1} + \frac{1}{1836} = 1.000545$$

$$\mu = 0.9994556$$

For hydrogen: $\lambda = 6560/0.9994556 = 6563.57$ Å

For deuterium: $\lambda = 6560/0.9997277 = 6561.79$ Å

The difference between these two values is 1.78 Å.

Example 3.7. It is possible to form a hydrogen-like atom called positronium in which the antiparticle of the electron, the positron, replaces the proton. The mass and charge of the positron are the same as for the electron except that the charge on the positron is positive. Calculate the energy required to ionize such an atom knowing that the energy required for hydrogen is 13.6 ev. What can be said about the spectral lines from such an atom as compared with hydrogen?

Solution. The reduced mass for positronium will be $1/\mu = 1/1 + 1/1 = 2$ so that μ = one-half the electron mass. This means that the energies for the various levels as given by equation 3.6 and the energy differences as reflected in equation 3.7 will be reduced by a factor of two, so that the ionization energy for positronium will be $13.6 \div 2$ or about 6.8 ev. Since the energy difference between any pair of levels will be reduced also by a factor of two, all wavelengths will be greater by a factor of two than the corresponding wavelengths for ordinary hydrogen.

Example 3.8. Calculate the radius of the first Bohr orbit for singly-ionized helium.

Solution. Using equation 3.4, we note that r depends inversely upon the charge of the nucleus. Thus, if $r_1 = 0.529$ Å for hydrogen, its value for singly-ionized helium, with a nuclear charge of 2e, should be half this value, or 0.264 Å.

Example 3.9. It is possible for a μ-meson to be captured in a Bohr orbit in an atom. A μ-meson is identical to an electron except that its mass is about 200 times larger than the electron. Calculate the radius of the first Bohr orbit for a μ-meson captured by a lead nucleus ($Z = 82$). The meson orbit is so close to the nucleus that one can assume that the orbital electrons have very little effect, so the calculation can be made as if they were absent. Recall that the first Bohr orbit for hydrogen has a radius of 0.529 Å.

Solution. Since $r = n^2h^2/4\pi mZe^2$, we expect r for this mu-mesic atom to be smaller by a factor or $1/82 \times 1/200$, so that:

$$r = \frac{0.592}{82 \times 200} = 3.61 \times 10^{-5}\text{ Å} = 3.61 \times 10^{-13}\text{ cm}$$

According to the best measurements of nuclear radii, this would place the meson somewhat inside the nucleus.

We have simplified our calculation by assuming that the nucleus was a point charge. On this assumption, it is calculated that the energy difference between the first and second Bohr orbits should be about 16.4 Mev. Energetic photons corresponding to a transition between these two levels have been observed, and these have an energy of about 6 Mev. By using this information and calculations of the effect of the nuclear charge being spread uniformly through a spherical nucleus, it has been possible to estimate nuclear radii for many elements.

3-4. Other Evidence For Quantized Energy States. The observation of discrete line spectra from all elements and the marked success of the Bohr model are strong evidence that the energy states in atoms are quantized. It is nàtural, however, to ask if the transfer of energy in atomic systems by any mechanism occurs in quantized or discrete amounts which are related to the wavelengths of observed spectral lines by the equation $E = h\nu = hc/\lambda$.

In 1914 the German scientists Franck and Hertz performed experiments in which atoms in the gaseous state were bombarded with electrons of known energy. Their idea was to see if these electrons could excite the atoms to higher states, if the energy given to the atoms always had the same quantized values, and if these energy values agreed with those determined by spectroscopic measurements.

An experimental arrangement, shown schematically in Figure 3.3, was used in which electrons, evaporated from a hot filament

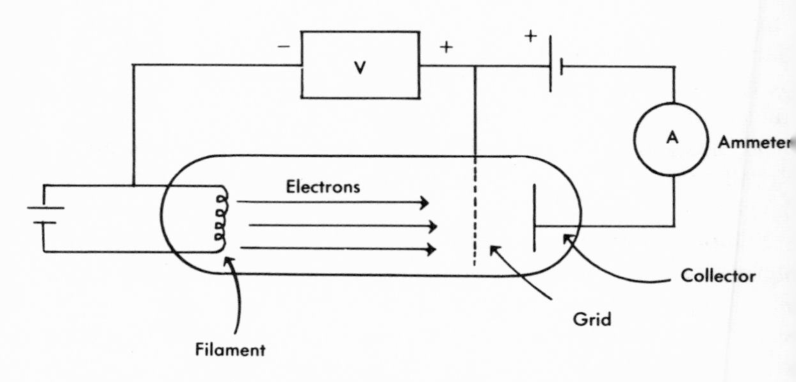

Figure 3.3. Experimental arrangement for the Franck-Hertz experiment.

were accelerated through a known potential difference applied between the filament and the grid. The tube is filled with a gas or vapor at low pressure, and the grid is a wire mesh through which the electrons can pass. Monatomic materials such as mercury neon, and argon are elements which have been used in such experiments to avoid the possibility of molecular excitations. The density of atoms in the tube must be such that the electrons have a reasonable probability of reaching the grid without too many collisions. In the case of mercury, this can be done by heating the evacuated tube so as to evaporate more of the mercury, normally present as a small liquid drop. As the voltage between filament and grid is increased, the current rises and then drops sharply. As the voltage is increased further, a series of such peaks can be observed as shown in Figure 3.4.

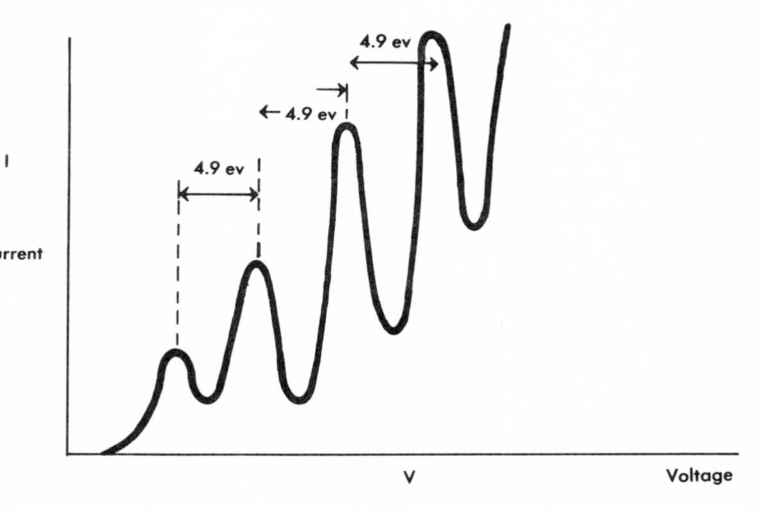

Figure 3.4. Current vs. voltage curve for Franck-Hertz experiment.

The interpretation of this curve is that when the kinetic energy of the electrons reaches a value just equal to the first excitation potential, which for mercury is 4.86 ev, inelastic collisions occur in which the electron gives all of its energy to the atom, exciting one of the outer or valence electrons to its first excited state. Momentum must be conserved in such a collision, but the recoil energy of the atom required by momentum conservation is negligible. Note in Figure 3.3 that a retarding potential of about one volt is applied between the grid and the collector of the vacuum tube. This potential prevents the electrons which have undergone inelastic collisions near the grid from being collected and produces the dips or peaks of Figure 3.4. The collisions producing the first peak must have occurred very close to the grid. As the voltage is increased, the electrons eventually will have sufficient energy to cause inelastic collisions about halfway from filament to grid. They then regain energy and repeat the process at or near the grid, thus producing electrons which cannot reach the collector causing the second peak.

As many as twenty such peaks can be observed if proper care is taken with the adjustment of the vapor pressure in the tube and with the electrical measurements. In the case of mercury, it is found that the distance between peaks is always 4.86 volts, within experimental error, which corresponds to a strong line in the mercury spectrum at 2537 Å. Furthermore, when the tube is observed with a spectrometer, light of wavelength 2537 Å is observed when the voltage accelerating the electrons is 4.9 volts or higher, but not for values less than 4.9 volts. Similar results are obtained for other elements showing that energy transfers are quantized for electron collisions and strongly indicating that such is the case for all mechanisms of energy transfer in atoms.

Example 3.10. Calculate the energy, in electron volts, corresponding to the transition in the mercury atom having a wavelength of 2537 Å.

Solution. The energy difference is given by:

$$E = h\nu = \frac{hc}{\lambda}$$

Recall that $h = 6.626 \times 10^{-34}$ joule-sec, $c = 3 \times 10^{8}$ m/sec, and 1 ev $= 1.6 \times 10^{-19}$ joules.

$$E = 6.625 \times 10^{-34} \times \frac{3 \times 10^{8}\ \text{m/sec}}{2537 \times 10^{-10} \times 1.6 \times 10^{-19}}$$

$$= 4.89\ \text{ev}$$

3-5. Resonance Absorption in Atoms. One might also expect that a photon having just the right energy might be capable of giving this energy to an atom and exciting the atom to an energy state higher by an amount exactly equal to the photon energy. This is found to be the case, and can be observed by passing white light, which contains all wavelengths in the visible region, through a gas or vapor. Many lines observed in the emission spectrum of an element are thus observed as dark lines in what is called an absorption spectrum. A photon of the proper frequency encountering an atom in the gas can excite it, for example, to its first excited state. The atom will remain in this state for a very short time and then give off a photon of the same frequency as the one it absorbed as it falls back to the ground state. This photon, however, will almost certainly be emitted in a different direction from that of the original photon, with the effect that light at this characteristic wavelength is removed from the beam passing through the sample. If many atoms absorb in this manner, the result is a measurable decrease in intensity at this wavelength, producing one of the dark lines in the absorption spectrum. Of course, if the atom is excited to a state higher than the first, it can reradiate more than one wavelength as it falls back to the ground state in several steps.

Absorption lines are observed by astronomers in the light coming from the sun as well as from other more distant stars. The positions of these lines, and to some extent, their relative intensities provide information about the elements which are present in the cooler outer layers of the sun's atmosphere and their relative abundances.

In laboratory experiments with resonance absorption, it is possible to observe the reradiated light at the expected frequency, as was the case with the Franck-Hertz experiment. This light is called resonance radiation.

It is interesting to note that this effect has been observed for atomic systems for many years, the first experiments having been done by the American physicist R. W. Wood in 1904 using sodium vapor. It turns out that the nuclei of atoms can also exist in quantized excited states much like those for the atom. For both atomic and nuclear states, the lifetime and the width (in energy) are related by the uncertainty principle so that $\Delta E \Delta t \approx h$. For atomic states, the lifetime is very short, being of the order of 10^{-7} or 10^{-8} seconds. For nuclear states, the lifetimes can vary over an extremely wide range from about 10^{-14} seconds to 10^{18} seconds. The problem arises because the "nuclear photons," or gamma rays, are much more energetic, often being of the order of one Mev (10^6 ev).

Since momentum must be conserved when the gamma ray is emitted, the nucleus recoils and in so doing acquires a certain energy which means that the photon is reduced in energy by this amount. Similarly, in being absorbed, a gamma ray causes the absorbing nucleus to recoil again, reducing the energy available to excite the nucleus. If these two processes result in an energy shift for the photon so great that resonance is spoiled (i.e., the shift is greater than the width of the excited state), then no absorption is possible.

Let us calculate the width of a state with a lifetime of 10^{-7} sec and see what the recoil energies are for a 2.5 ev photon (λ = 5000 Å) and a 100,000 ev gamma ray.

$$\Delta E = \frac{h}{\Delta t} = \frac{6.625 \times 10^{-34}}{10^{-7}} = 6.625 \times 10^{-27} \text{ joules}$$

Dividing by 1.6×10^{-19} joules per ev gives about 4×10^{-8} ev for the width of the level.

It is relatively simple to calculate the energy of the recoiling atom of mass m. We equate the momentum of the gamma ray or photon to that of the recoiling atom; i.e., $mv = h\nu/c$. Squaring both sides gives $m^2v^2 = E_\gamma^2/c^2$, where $h\nu = E_\gamma$. Thus $\frac{1}{2}mv^2 = E_\gamma^2/2mc^2 = E_R$, where E_R is the recoil energy. For a photon of wavelength 5000 Å and energy about 2.5 ev, we get for an arbitrary atomic weight of 100:

$$E_R = \frac{2.5^2}{2 \times 100 \times 931 \times 10^6} = 0.33 \times 10^{-10} \text{ ev}$$

which is small compared to a typical level width.

On the other hand, if we make the same calculation for a 100,000 ev gamma ray, we get $E_R = 10^{10} \div (2 \times 100 \times 931 \times 10^6) = 0.054$ ev—which is considerably larger than the level width but only about five parts in 10^7 of the gamma ray energy. Note that in the above calculation we have utilized the fact that one amu is equal to 931 Mev, and have expressed both E_γ and mc^2 in units of electron volts.

In 1958, R. L. Mossbauer devised techniques to permit resonance absorption of gamma rays from nuclei. The idea is to use (for obvious reasons) fairly low-energy photons (perhaps 15,000 ev) and to use sources and absorbers in the form of crystals so that, instead of a single atom recoiling, the whole crystal must recoil because the atom is firmly bound in position in the lattice. Thus, the recoiling mass is essentially infinite and the energy shift experienced by the photon practically zero. The photon energy should also be sufficiently low that not many vibrational motions in the crystal are started by the recoiling atom. The energies of such vibrational motions in crystals are quantized and are called *phonons*. The phonon might be thought of as a quantized bundle of sound energy. No phonons can be produced until the energy available is equal to or greater than that required to excite the first, or lowest, excitation.

One of the best nuclei for Mossbauer experiments is ^{57}Co which results from the radioactive decay of ^{57}Fe. One of the photons, or gamma rays, from this nucleus has an energy of 14.4 Kev, and the excited state from which it originates has a lifetime of about 10^{-7} seconds. This means that the width of this excited state is about 5×10^{-9} ev.

The great sensitivity of the Mossbauer technique stems from the fact that if the source is moved relative to an absorber with a velocity of a few tenths of a millimeter per second, resonance absorption can be spoiled because of the Doppler effect, which changes the frequency of the photon very slightly. The sensitivity is measured by the ratio of the level width to the photon energy, this ratio being about one part in 10^{12} for ^{57}Fe. This means that if for any reason the energy of the photon is changed (i.e., the position of the level in the nucleus is changed slightly), thus spoiling the resonance absorption, this change can be compensated (and measured) by moving the source carefully at a known velocity. When nuclei are put in a magnetic field very slight changes occur in the energy levels. This is known as the Zeeman effect and occurs also for atomic energy levels. The magnetic field seen by an atom in different parts of a molecule can be slightly different and these slight differences can be measured using the Mossbauer technique.

The almost incredible sensitivity of the Mossbauer effect has also permitted the direct verification of one of the predictions of the General Theory of Relativity. This prediction is that the frequency of a photon emitted in a strong gravitational field will be lower than that of one emitted in a weaker gravitational field. R. V. Pound and G. A. Rebka, Jr. measured this effect by separating ^{57}Fe sources and absorbers vertically by some 74 feet. They observed a fractional frequency shift of 5.13×10^{-15} which agreed within 5% with the predicted value from the Theory of Relativity.

Chapter 4
Atoms With More Than One Electron

4-1. Introduction. Despite the success of the Bohr theory and its attractive simplicity, it must be emphasized that it is a useful oversimplification which permits us to think about hydrogen-like atoms using many familiar classical ideas. The Bohr theory provides no means of calculating the relative probabilities of the various transitions, and hence the relative intensities of spectral lines. It does not predict the very slight splitting of lines known as fine structure. It cannot handle many electron atoms, and it provides no explanation of the forces which hold atoms together in molecules or in the solid or liquid state.

All of these deficiencies can be remedied by a wave-mechanical treatment of considerably greater complexity, based upon the wave nature of matter. In terms of wave mechanics it makes no sense to talk in terms of particles of a definite size moving in clear-cut orbits. One must think in terms of a wave function usually a sinusoidal or exponential function, which contains all the information about the system. The square of this wave function gives a measure of the probability of finding an electron at a particular place in the atom. The wave function is a solution of a second-order differential equation, known as the *Schroedinger equation*, which forms the basis of quantum mechanics when formulated in terms of the de Broglie hypothesis. This equation provides a connection between the de Broglie waves and the particle with which they are somehow connected. Solutions of this equation provide the wave functions which have been used successfully to calculate the properties of atomic and molecular systems; although for atoms with many electrons, the calculations become very involved.

For the hydrogen atom, the various possible solutions would correspond to the Bohr orbits we have already discussed, except that the information from wave mechanics is provided in terms of probabilities (indicated in Figure 4.1) in which the probability of finding the electron at a distance r from the nucleus is plotted versus r. The maximum of this curve comes at exactly the value of r predicted by the Bohr theory, but we note that wave

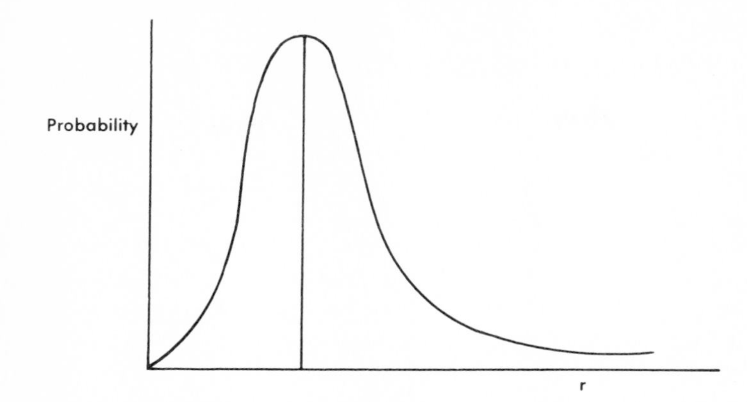

Figure 4.1. Probability of finding an electron at a distance r from the nucleus for the ground state of the hydrogen atom.

mechanics predicts that the electron spends considerable time at other radii.

4-2. Quantum Numbers for Atomic Systems. The Bohr treatment of the hydrogen atom involved only one quantum number, which we called n. When the Schroedinger equation is solved for atomic systems with relativistic effects taken into account, four quantum numbers arise. We shall discuss each of these qualitatively, emphasizing their physical significance or interpretation, and then use them to discuss the electronic structure of the atoms formed as more and more electrons are added to "build up" the Periodic Table. It should be pointed out that exact solutions are not possible for atomic systems made up of more than two bodies but approximation techniques can be used for many electron atoms which give good agreement with experiment.

4-3. The Orbital Angular Momentum Quantum Number. Wave mechanics predicts that the orbital angular momentum of an electron in an atom is quantized and has the value $P_l = \sqrt{l(l + 1)}\hbar$, where l is called the orbital angular momentum quantum number and $\hbar$ is Planck's constant divided by 2π. The possible values of l are related to the principal quantum number n, similar to the n used in the Bohr theory, and can have values of $l = 0, 1, 2 \ldots (n - 1)$. Classically, we associate the angular momentum with the shape of an orbit. The more circular the orbit, the larger the angular momentum, which would be equal to mvr for a circular orbit.

Classically, $l = 0$ would correspond to an orbit in the form of a straight line. This would mean that an $l = 0$ electron would have to pass right through the nucleus which is classically unacceptable, but in terms of wave mechanics presents no problem since in this formalism one deals only with wave functions and not with mechanical models. The wave functions for $l = 0$ electrons are spherically symmetrical—but such is not the case for higher l values. Thus the value of the orbital angular momentum quantum number determines the shape of electron orbits, or more properly determines the probability distribution or the shape of what might be called the electron density function for an electron. The shape of the $l = 1$ orbit is roughly that of the three-dimensional equivalent of a figure eight.

Information about the shape of electron probability distributions has been of considerable help to chemists in explaining the forces which hold atoms in molecules and in explaining some aspects of chemical reactions. It has become conventional to refer to the values of l using letters as well as numbers, so that the l values 0, 1, 2, 3, and 4 are often represented by the letters *S*, *P*, *D*, *F*, and *G*, respectively.

4-4. The Magnetic Quantum Number m_l. A third quantum number resulting from the wave-mechanical approach to atomic systems is called the magnetic quantum number and can be represented by m_l. We interpret this physically in terms of a phenomena known as space quantization. Angular momentum is a vector quantity and, therefore, the angular momentum of an electron can be considered to have a certain direction in space relative to any reference direction such as that of an external magnetic field. The angular momentum of an object moving in a circular path is perpendicular to the plane of the loop and follows a right hand rule so that if the fingers of the right hand curl in the direction of motion, the thumb points in the direction of the angular momentum. Similarly, a current loop, which behaves much like a bar magnet, is said to have a magnetic moment of magnitude IA, where I is the current and A is the area of the loop (also directed perpendicular to the loop). Classically, we know that current loops in a magnetic field experience a torque so that the area of the loop tends to become perpendicular to the field and the magnetic moment parallel to it. Any orientation is possible but work is required to orient the magnetic moment so that it makes an angle Θ with the field.

Experimental evidence, which we discuss below, indicates that for electrons in atoms the orientation of the magnetic moment and, therefore, that of the angular momentum are quantized, i.e., only certain spatial orientations are permitted. This is illustrated in Figure 4.2 for an electron with $l = 3$, i.e., an F state. Quantum

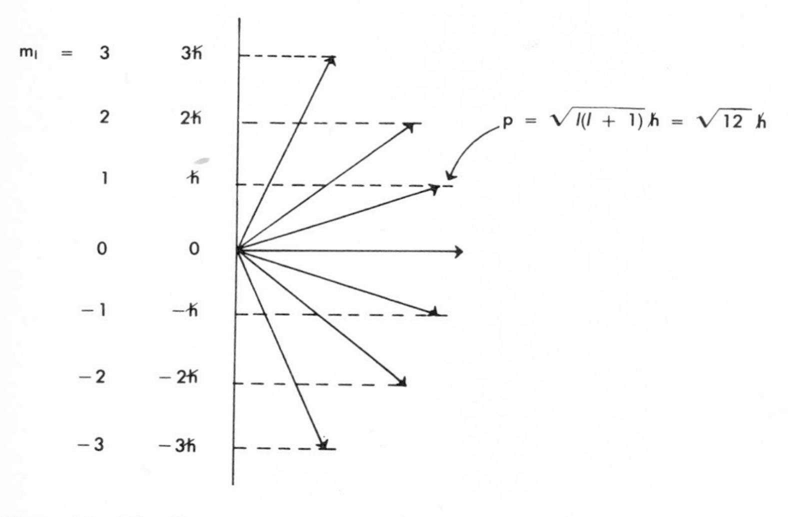

Figure 4.2. The allowed spatial orientations of the orbital angular momentum vector for an F state ($l = 3$).

mechanics requires that the projection of the orbital angular momentum vector on the direction of an external magnetic field be limited to integral values of $h/2\pi$. It also predicts that the magnitude of the angular momentum is given by $\sqrt{l(l+1)}$ so that, as illustrated in Figure 4.2, the angular momentum is never parallel to the field. The magnetic quantum number m_l varies from $+l$ to $-l$ and includes zero so that there are $2l + 1$ possible orientations. If this is the case, and each orientation represents a different energy as we have implied, the energy levels in atoms placed in a magnetic field should be split, more transitions should be possible, and more lines should be observed. Since the splitting of energy levels is small, this effect is revealed by a splitting of the spectral lines into several components very close in wavelength. This effect is known as the *Zeemann effect*, and is detectable only with very sensitive high-resolution spectroscopic equipment. The predictions of quantum mechanics agree well with experiment when the effects of electron spin, to be discussed shortly, are included.

4-5. Electron Spin. The quantum numbers n, l, and m_l discussed above arise naturally from a nonrelativistic treatment of atomic systems using wave mechanics. This treatment, however, was unable to explain the Zeeman splitting observed for most atoms and was also unable to explain the fact that most spectral lines were split into two lines separated by roughly an Angstrom unit.

It was suggested in 1925 by the Dutch scientists Goudsmit and Uhlenbeck that the electron behaves as though it were spinning, much as the earth rotates on its axis and, therefore, should have a spin angular momentum and a spin magnetic moment in addition to its orbital angular momentum and magnetic moment. Fine structure and the so-called analogous Zeeman effect could be explained if it was assumed that the electron was assigned a spin quantum number s which could have only the values $+\frac{1}{2}$ and $-\frac{1}{2}$. This reflects the fact that the spin angular momentum, given by $P_s = \sqrt{s(s + 1)}$, can have only two orientations in space such that its projection on the direction of a magnetic field is either $+\frac{1}{2}h/2\pi$ or $-\frac{1}{2}h/2\pi$. The fine structure in atomic spectra can be considered as an internal Zeeman effect or spin-orbit interaction in which the energy of an electron is slightly different depending upon whether its spin is "up" or "down" relative to the internal magnetic field in the atom produced by its orbital motion and by other electrons.

In 1928, P.A.M. Dirac was successful in obtaining a relativistic solution of the Schroedinger equation, which predicted electron spin as well as the other three quantum numbers, eliminating the necessity for its introduction on a purely empirical basis.

4-6. The Stern-Gerlach Experiment. Although the agreement of theory and experiment for the Zeeman effect can be considered a verification of the concept of space quantization for orbital and spin angular momentum, a somewhat more direct and convincing experiment was performed in 1921 by the German scientists Stern and Gerlach. Their experimental arrangement is shown schematically in Figure 4.3.

The basic idea of the experiment is to pass a beam of neutral atoms through an extremely nonuniform magnetic field. In the

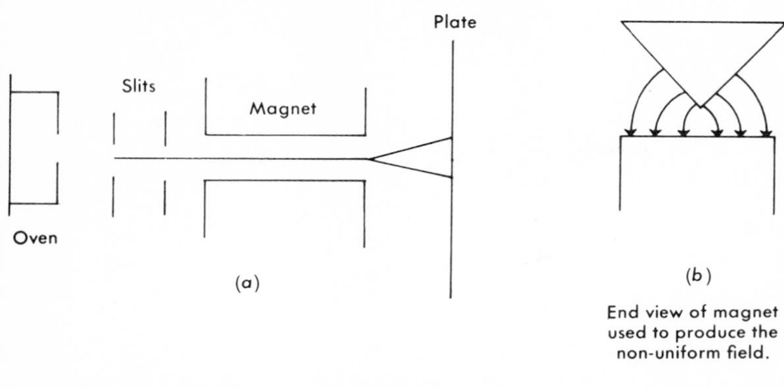

Figure 4.3. Experimental arrangement for the Stern-Gerlach experiment.

field, the atoms may have various possible space orientations. For simplicity, we might imagine each atom as a small bar magnet which normally would experience no net force in a magnetic field. If the field at one end is sufficiently different from that at the other, there will be a net deflecting force which will depend on the space orientation of the "magnet." Stern and Gerlach used silver atoms which came from an oven in which the silver was evaporated and existed as a gas. Instead of producing a smear on the photographic plate (as would be the case if all orientations were possible) or a straight line (as would be the case if there were no deflection or all atoms experienced the same deflection), the experiment produced two lines. This was direct evidence for space quantization, but was not in agreement with wave mechanical predictions at the time since electron spin was not "discovered" until about 1925. From our discussion of the space quantization of the orbital angular momentum we would expect one line on the plate corresponding to an angular momentum of zero, three lines corresponding to $l = 1$, or more than three for higher l values, but in no case two lines. This difficulty was resolved after the discovery of electron spin when it was realized that the entire magnetic moment of the silver atom is due to a single electron, the combined effects of the others exactly canceling because of the shell structure of the electrons (which we will discuss in the next section).

Thus, one would expect that as the silver atoms entered the magnetic field, they would align themselves either parallel or antiparallel to the field and would therefore be deflected either up or down in Figure 4.3 if the field were sufficiently nonuniform, producing a pattern on the plate with only two lines as was observed.

4-7. The Periodic Table and the Pauli Exclusion Principle. In the preceding sections we have discussed the four quantum numbers, n, l, m_l, and s to which we gave the following general significance. The principal quantum number, as in the Bohr model, is a rough measure of the binding energy of the electron in the atom and a measure of the average distance from the nucleus, although this depends also on the value of l in many cases. The orbital angular momentum quantum number, l, determines the shape of the orbit and determines the number of possible space orientations for the angular momentum vector, each orientation corresponding to a different value of m_l, the magnetic quantum number. Finally, the spin quantum number, s, represents the fact that a certain angular momentum is associated with the electron itself (as though it were spinning). There are two possible values of s, plus or minus one-half, indicating that the spin angular momentum has only two possible spatial orientations.

The *Pauli exclusion principle*, proposed by W. Pauli in 1925, says that no two electrons in any one atom may have the same values for all four of the above quantum numbers.

Quantum mechanical calculations of the properties of many-electron atoms are possible in principle, but extremely complicated in practice. Using the exclusion principle in a very simple way we can, however, gain considerable insight about the chemical periodicities observed as the result of what has come to be known as the shell structure of the atomic electrons. We start with hydrogen and move from one value of the atomic number Z to the next by adding a single electron and assuming that the nuclear charge also increases by one. We assume, as confirmed by experiment, that the lowest possible values of n and l are those which the electrons take on first. We start with $n = 1$ and ask how many electrons we can add before we can add no more without violating the Pauli exclusion principle. This process is illustrated in Table 4-1, in which we associate with each element all the electrons listed for elements of lower Z.

When $n = 1$, we recall that l can only be zero, hence m_l must be zero so that only two electrons are permitted with $n = 1$: one corresponding to $s = +\frac{1}{2}$ the other to $s = -\frac{1}{2}$. Hydrogen has

TABLE 4-1. Illustrating the manner in which the Pauli exclusion principle is used in assigning quantum numbers to electrons in atoms. Note that nitrogen, for example, has six electrons with the same quantum numbers as the six electrons in carbon, while the seventh has the quantum numbers as listed opposite nitrogen in the table.

Z	Element	n	l	m_l	s
K shell: 1	H	1	0	0	$+\frac{1}{2}$
K shell: 2	He	1	0	0	$-\frac{1}{2}$
L shell: 3	Li	2	0	0	$+\frac{1}{2}$
L shell: 4	Be	2	0	0	$-\frac{1}{2}$
L shell: 5	B	2	1	-1	$+\frac{1}{2}$
L shell: 6	C	2	1	-1	$-\frac{1}{2}$
L shell: 7	N	2	1	0	$+\frac{1}{2}$
L shell: 8	O	2	1	0	$-\frac{1}{2}$
L shell: 9	F	2	1	$+1$	$+\frac{1}{2}$
L shell: 10	Ne	2	1	$+1$	$-\frac{1}{2}$
11	Na	3	0	0	$+\frac{1}{2}$

one electron with $n = 1$ and $l = 0$, and helium has two. Lithium, however, which has three electrons, must have one of them with $n = 2$ and $l = 0$. Thus, the first shell is said to close at helium and lithium has a single electron which starts the second shell. Traditionally, the letters *K*, *L*, *M*, and *N* (etc.) are used to represent the shells corresponding to $n = 1, 2, 3$, and 4, etc. The *L* shell ($n = 2$) has "room," according to the Pauli principle, for many more electrons. For this shell, when $n = 2$, l may be either zero or one, and when $l = 1$, m_l may be either -1, 0, or $+1$ so that (as indicated in Table 4-1) the *L* shell, with a total of eight electrons, is filled at Z = 10 (neon). Two of the eight *L* electrons have $l = 0$, and the other six are *p*-electrons ($l = 1$). Thus, each shell contains $2n^2$ electrons, and each subshell, corresponding to the different l values with a shell, contains $2(2l + 1)$ electrons.

Table 4-2 indicates the manner in which this process continues through the Periodic Table up to Z = 41 using the generally accepted spectroscopic notation. Shown also is the ionization potential for each element, providing a good example of the kind of periodicity produced by this shell structure. Some exceptions

TABLE 4-2. Electronic shell structure through Z = 41.

Element		*K*	*L*		*M*			*N*				Ionization
	Z	1*s*	2*s*	2*p*	3*s*	3*p*	3*d*	4*s*	4*p*	4*d*	4*f*	Potential
H	1	1										13.6 ev
He	2	2										24.5
Li	3	2	1									5.4
Be	4	2	2									9.3
B	5	2	2	1								8.2
C	6	2	2	2								11.2
N	7	2	2	3								14.5
O	8	2	2	4								13.6
F	9	2	2	5								18.6
Ne	10	2	2	6								21.5
Na	11	2	2	6	1							5.1
Mg	12	2	2	6	2							7.6
Al	13	2	2	6	2	1						6.0
Si	14	2	2	6	2	2						8.1
P	15	2	2	6	2	3						11.1
S	16	2	2	6	2	4						10.3
Cl	17	2	2	6	2	5						13.0
A	18	2	2	6	2	6						15.7
K	19	2	2	6	2	6		1				4.3
Ca	20	2	2	6	2	6		2				6.1
Sc	21	2	2	6	2	6	1	2				6.7
Ti	22	2	2	6	2	6	2	2				6.8
V	23	2	2	6	2	6	3	2				6.8
Cr	24	2	2	6	2	6	5	1				6.7
Mn	25	2	2	6	2	6	5	2				7.4
Fe	26	2	2	6	2	6	6	2				7.8
Co	27	2	2	6	2	6	7	2				8.5
Ni	28	2	2	6	2	6	8	2				7.6
Cu	29	2	2	6	2	6	10	1				7.7
Zn	30	2	2	6	2	6	10	2				9.4
Ga	31	2	2	6	2	6	10	2	1			6.0
Ge	32	2	2	6	2	6	10	2	2			8.1
As	33	2	2	6	2	6	10	2	3			10.5
Se	34	2	2	6	2	6	10	2	4			9.7
Br	35	2	2	6	2	6	10	2	5			11.3
Kr	36	2	2	6	2	6	10	2	6		5*s*	13.9
Rb	37	2	2	6	2	6	10	2	6		1	4.2
Sr	38	2	2	6	2	6	10	2	6		2	5.7
Y	39	2	2	6	2	6	10	2	6	1	2	6.5
Zr	40	2	2	6	2	6	10	2	6	2	2	6.9
Nb	41	2	2	6	2	6	10	2	6	4	1	6.8

to the rules we have outlined appear, the first being at potassium ($Z = 19$), where we see that a $4s$ electron ($n = 4$, $l = 0$) has a lower energy, i.e., is more tightly bound to the atom, than a $3d$ electron. Therefore, potassium in its lowest, or ground, state has its outermost electron in the N shell instead of the M shell. A similar situation exists at $_{37}$Rb where a $5s$ electron shows up instead of a $4d$ electron. This can be explained qualitatively by recalling that s electrons ($l = 0$) would correspond classically to a straight-line orbit going through the nucleus. Quantum mechanically it can be said that they have a large probability of getting fairly close to the nucleus, that they spend more time near the nucleus than do electrons with higher l values, and are thus shielded less by other electrons and are more strongly attracted to the nucleus and hence are more tightly bound.

Often the electronic configuration of an atom is represented using the spectroscopic notation used in Table 4-2. As an example, consider the sodium atom. The ground state configuration for sodium is given by $1s^2$, $2s^2$, $2p^6$, $3s^1$ where the first number gives the n value, the letter gives the l value, and the superscript gives the number of electrons with a particular l and n value—i.e., $2p^6$ indicates that there are six electrons with $n = 2$ and $l = 1$.

It is the electrons in the unfilled outermost shell which determine the chemical properties of the various elements. Lithium, sodium, and potassium, for example, with a valence of one are very active chemically because the single electron in the outermost shell is removed relatively easily, facilitating many chemical reactions. Elements such as helium, neon, and argon (the rare gases), with outer shells completely filled, are virtually inert chemically.

The chemical similarity of elements in the same column of the Periodic Table such as boron, aluminum, gallium, and indium results because they have the same number of electrons in the outermost shell or subshell, and these electrons all have the same l value. For the elements just mentioned, the outer electron configurations are $2p^1$, $3p^1$, $4p^1$, and $5p^1$, respectively. For fluorine, chlorine, bromine, and iodine, the so-called halogens, the configurations are $2p^5$, $3p^5$, $4p^5$, and $5p^5$, respectively, indicating that these elements are all one electron short of a closed shell. Because of this configuration, they are chemically active because they "like" to accept electrons to form a closed shell.

It can be shown that the total angular momentum and magnetic moment of all the electrons forming closed or filled shells is zero. Thus, the single valence electron in the silver atom used in the Stern-Gerlach experiment is a 5*s* electron, indicating why the total angular momentum of this atom is due only to this electron and why its magnetic moment is that of a single electron.

The optical spectrum is also due only to the outer, or valence, electrons in atoms. In general, the spectrum for a particular atom is quite complex, but for certain elements, such as sodium, the spectrum should be quite similar to hydrogen since a single electron moves about the nucleus which is shielded by the other electrons which form closed shells. It turns out that the spectrum is similar but more complicated because, in sodium, levels having the same n value but different l values have different energies; for hydrogen the energy was independent of l.

Example 4.1. Give the ground state electronic configuration for germanium.

Solution. From Table 4-2, we see that the 32 electrons in germanium must be distributed as follows:

$$1s^2, 2s^2, 2p^6, 3s^2, 3p^6, 3d^{10}, 4s^2, 4p^2$$

Example 4.2. Consider the sodium atom in its ground state. What are the n and l values of the first two excited states of the 3*s* electron, which is alone in the M shell?

Solution. From Table 4-2, we see that the first excited state should be a 3*p* state, corresponding to $n = 3$ and $l = 1$, and that the second excited state should be a 4*s* state as indicated by the irregularity at potassium.

Example 4.3. The ionization potential for $_3$Li is given as 5.4 ev in Table 4-2. Assuming that lithium can be represented as a single electron moving about the closed K shell and nucleus with a net positive charge of $3 - 2 = 1$ electronic charge, what would you expect the ionization potential to be, and how can one qualitatively account for the difference?

Solution. Since the exclusion principle requires that this outer electron have a principal quantum number of 2, one would expect that if the effective nuclear charge were indeed one, then the ionization energy would be the same as for the first excited state of hydrogen which would be $13.6 - 10.2 = 3.4$ ev, as indicated in Figure 3.1. However, since an *s* electron has a relatively high probability of being close to the nucleus, we can explain the larger experimental value by pointing out that the 2*s* electron probably

enetrates inside the K shell on occasion so that the effective
nuclear charge is greater than one, and the binding energy is
orrespondingly larger.

4-8. *X* Rays. In 1895, while conducting experiments with
athode ray tubes similar to those discussed in Chapter 1, the
erman physicist W. C. Roentgen (1845–1923) discovered that a
ery penetrating radiation came from his tube which could be
etected on a fluorescent screen some distance from the tube.
oentgen found that this radiation, which has since been called
X radiation, traveled in straight lines and was not affected by
nagnetic fields. He was able to make the first crude *X*-ray photo-
raphs showing the bones in a human hand.

We now know that these *X* rays, which typically have wave-
engths of the order of 1 Å, are energetic photons which are pro-
uced whenever high energy electrons strike a target of any mate-
ial. A typical *X*-ray tube is sketched in Figure 4.4. Usually, metal

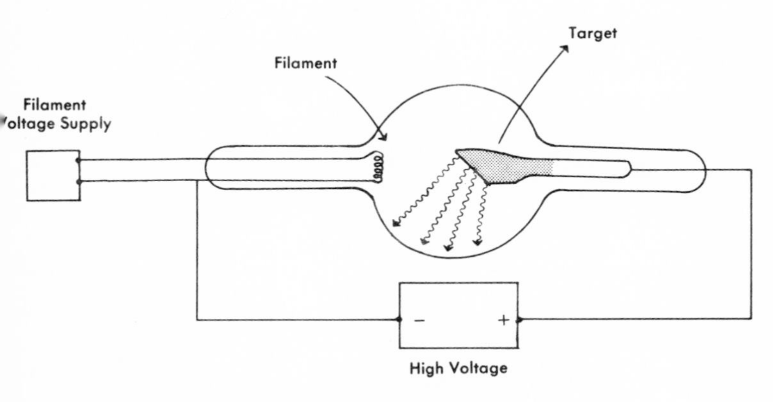

Figure 4.4. An *X*-ray tube.

argets are used, copper and molybdenum being common mate-
ials, with the electrons being accelerated through potential dif-
erences in the range of 50,000 to 100,000 volts.

It is found that two types of *X* rays are produced, as indicated
n Figure 4.5 which shows a typical *X*-ray spectrum. Super-
nposed on a continuous background are sharp peaks at particu-
ar wavelengths which are different for different target materials.
n the following sections, we will discuss the origin of both the
ontinuous and so-called characteristic, or sharp line, *X*-ray spec-

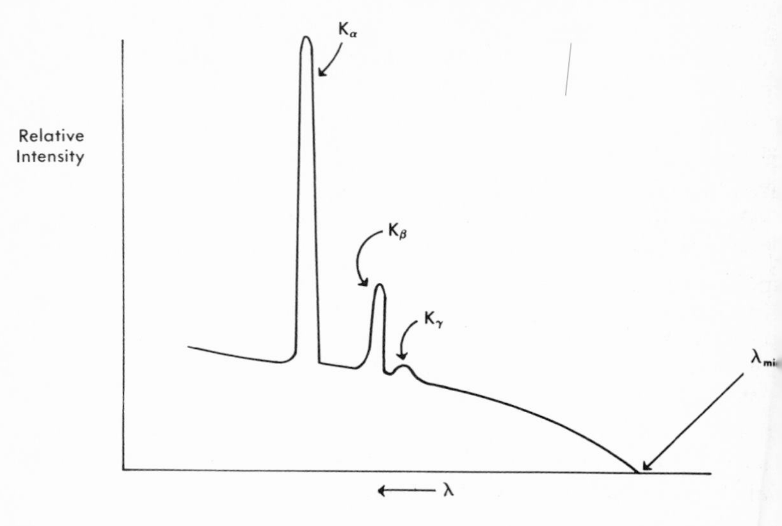

Figure 4.5. An X-ray spectrum showing both characteristic and continuou X-ray spectra. Here only the K series is shown. The other serie would be at longer wavelengths.

tra, as well as the techniques which have been developed for the measurement of this very short wavelength radiation.

4-9. The Continuous *X*-Ray Spectrum. The continuous X-ray spectrum is explained in terms of the sudden accelerations ex perienced in the target by the incident electrons as they are sud denly deflected and accelerated in the fields of target nuclei Electromagnetic theory predicts that when charges are acceler ated, they should radiate. Often the German word *Bremsstrahlung* (braking or retardation radiation) is applied to this process. The electrons oscillating back and forth in the antenna of a radio transmitter radiate radio waves for a similar reason, although the wavelength is considerably longer. It is also interesting to point out that electrons moving in circular paths in an accelerator called a synchrotron, at nearly constant velocity, are also ob served to radiate because they are constantly changing direction and hence are accelerated.

It is observed that the continuous X-ray spectrum goes up to a maximum energy $h\nu = hc/\lambda$ equal to the maximum kinetic energy of the incident electrons so that we may write:

$$\frac{hc}{\lambda} = Ve$$

and:

$$\lambda_{min} = \frac{hc}{Ve} \tag{4.1}$$

Equation 4.1 has been used as a method for measuring Planck's constant since the velocity of light and the charge on the electron are fairly well known, and everything else can be measured.

Example 4.4. Calculate Planck's constant from the information that the short wavelength limit of the continuous *X*-ray spectrum produced by 50 kilovolt electrons is 0.248 Å.

Solution. From Equation 4.1 we get:

$$\begin{aligned} h &= Ve\,\frac{\lambda}{c} \\ &= \frac{5 \times 10^{4} \times 1.6 \times 10^{-19} \times 0.248 \times 10^{-10}}{3 \times 10^{8}} \\ &= 6.61 \times 10^{-34} \text{ joule-sec} \end{aligned}$$

4-10. Characteristic *X*-Ray Spectra. Most of the electrons incident upon the target of an *X*-ray tube will encounter electrons in the outer shells. In order to transfer an appreciable amount of energy to such an electron, a very close encounter must occur. This is relatively rare because the outer electrons, in a manner of speaking, have more room and can be moved out of the way. Thus, most of the collisions result in energy transfers of the order of tens of electron volts. Although the inner shells take up a rather small portion of the atomic volume, it is possible for very energetic collisions to occur which knock inner shell electrons completely out of the atom or into unoccupied states in the unfilled outer shell. If this were to happen to a *K* electron in the innermost shell, an electron from one of the outer shells (say the *L* shell) could fall into this empty level and in the process give off an energetic photon. This would leave a vacancy in the *L* shell, which in turn could be filled by another electron from a higher energy state with the emission of another photon. This second photon would be lower in energy than the first. Thus, the vacancy produced by the original collision can work its way to the outermost shell of the atom, where it can be filled eventually by an electron knocked out of a neighboring atom.

These energetic photons are the characteristic *X* rays which appear superimposed on the continuous *X*-ray spectrum. Since the separation of levels is different in different atoms, we would expect these characteristic *X* rays to have different wavelengths for different elements. According to our discussion of the elec tronic structure of atoms, based on the Pauli exclusion principle we would expect, however, a certain similarity, regularity, and even, simplicity in the characteristic X-ray spectra because of th similarity of the inner shells in all atoms. For example, ther are only two electrons in the K shell for all atoms (except hydro gen), and all atoms above neon in the Periodic Table have eigh electrons in the L shell.

Figure 4.6 illustrates the transitions which produce the char acteristic X rays and gives the notation used in describing them

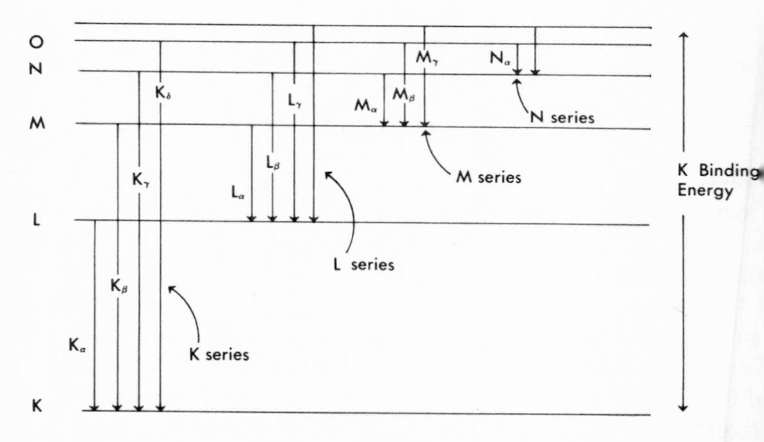

Figure 4.6. The electronic transitions which produce the characteristic X rays.

Note the great similarity to the transitions which occur in th hydrogen atom. This is not surprising since for the K electron in particular we would expect the situation to be much like th hydrogen atom since they are closest to the nucleus and no shielded by the other electrons.

Let us use the results of the Bohr model calculation, as give in equation 3.7, to calculate the wavelength of the K_α line fo copper, which has an atomic number of 29. We recall that fo hydrogen, as shown in Figure 3.1, the ionization potential is 13. ev. We can estimate the energy difference between the K and shells in copper by writing equation 3.7 as follows to give th energy difference between the ground state and the first excite state for a single electron moving about a nucleus of charge Ze.

$$\Delta E = 13.6Z^2\left(\frac{1}{1^2} - \frac{1}{2^2}\right) \tag{4.2}$$

If a K electron is knocked out of the copper atom, this leaves one electron in the K shell. It seems reasonable to assume that it is between the nucleus and the L electrons most of the time and, therefore, shields the nucleus so as to reduce the effective nuclear charge to $Z - 1$. Using the above equation, replacing Z^2 by $(Z - 1)^2$, and noting that the value of the parenthesis is 0.75 we get for copper: $\Delta E = 13.6 \times (28)^2 \times 0.75 = 7997$ ev. The measured value of the K_α line for copper is 1.54 Å which gives:

$$\Delta E = \frac{hc}{\lambda} = \frac{6.626 \times 10^{-34} \times 3 \times 10^8}{1.54 \times 10^{-10}}$$

$$= 12.9 \times 10^{-16}\,\text{joules} = \frac{12.9 \times 10^{-16}}{1.6 \times 10^{-19}}$$

$$= 8070\ \text{ev}$$

which is quite close.

Example 4.5. Calculate the wavelength of the K_α line from molybdenum ($Z = 42$) using the Bohr model and compare it with the measured value of 0.707 Å.

Solution. The energy difference between the K and L states in molybdenum can be calculated as above, using our previously calculated energy values for the hydrogen atom. We can write:

$$\Delta E = 13.6\,(Z - 1)^2\left(\frac{1}{1^2} - \frac{1}{2^2}\right)$$

$$= 13.6 \times (41)^2 \times 0.75 = 17{,}150\ \text{electron volts}$$

To calculate the wavelength, we use $\Delta E = hc/\lambda$ so that $\lambda = hc/\Delta E$ and:

$$\lambda = \frac{6.625 \times 10^{-34} \times 3 \times 10^8}{17{,}150 \times 1.6 \times 10^{-19}} = 0.724 \times 10^{-10}\ \text{m}$$

Example 4.6. A student in the laboratory is given an X-ray tube and told that the target is either iron ($Z = 26$) or copper ($Z = 29$). He measures the K_α line from this target and finds that its wavelength is 1.94 Å. Is the target made of iron or copper?

Solution. We assume that a simple Bohr model calculation will be adequate and write:

$$\Delta E = \frac{hc}{\lambda} = 13.6 \times (Z - 1)^2\left(\frac{1}{1^2} - \frac{1}{2^2}\right)$$

Solving for (Z − 1) we get:

$$(Z - 1)^2 = \frac{hc/\lambda \times 1}{13.6 \times 0.75 \times 1.6 \times 10^{-19}}$$

$$(Z - 1)^2 = \frac{6.625 \times 10^{-34} \times 3 \times 10^{8}}{1.94 \times 10^{-10} \times 13.6 \times 0.75 \times 1.6 \times 10^{-19}}$$

$$= 628$$

$$Z - 1 = 25$$

and Z = 26. Therefore, the target is iron.

It should be pointed out that the K series cannot be observe unless a K electron is removed. In order for this to be possible the accelerating voltage in the X-ray tube must be at least equa to the binding energy of the K electrons. Thus, for copper, th binding energy of the K electrons is 8.98 kilovolts. If the bindin energy of an L electron is 1000 volts, the K_α line will have a energy of 7.98 kilovolts, but it will not appear until the energy o the incident electrons is greater than 8980 volts.

Example 4.7. The binding energy of the L electrons in lead i about 15,000 volts. If the K_α line has a wavelength of 0.17 Å, wha is the minimum accelerating voltage in the X-ray tube which wil produce the K series?

Solution. The accelerating voltage must be the sum of the binding energy and the energy of the K_α line so that:

$$V = 15{,}000 + \frac{hc}{\lambda}$$

$$= 15{,}000 + \frac{6.625 \times 10^{-34} \times 3 \times 10^{8}}{0.17 \times 10^{-10} \times 1.6 \times 10^{-19}}$$

$$= 15{,}000 + 73{,}000 = 88{,}000 \text{ volts.}$$

We have oversimplified the situation somewhat because w have fine structure in X-ray spectra just as we do in optica spectra. Instead of one energy for the L electrons we actuall have three fairly close together. In the L shell there are two 2 electrons and six $2p$ electrons. Both the s electrons have the sam binding energy, but there are two possible energies for the electrons corresponding to the electron spin parallel and anti parallel to the orbital angular momentum. In lead, for example the binding energy of the two K electrons is 88,000 volts and th three binding energies for the L shell are 15,860, 15,200, an

13,000 volts. The situation is even more complicated in the *M* shell where there are five different binding energies. It is found that there are only two K_α lines and two K_β lines despite the fact that three energy states exist in the *L* shell and five in the *M* shell. The missing transitions are prohibited by selection rules provided by a quantum mechanical treatment.

4-11. The Measurement of *X*-Ray Wavelengths. As was the case with electron wavelengths, natural diffraction gratings in the form of crystalline materials are used in measuring *X*-ray wavelengths. Figure 4.7 shows schematically the makeup of an *X*-ray

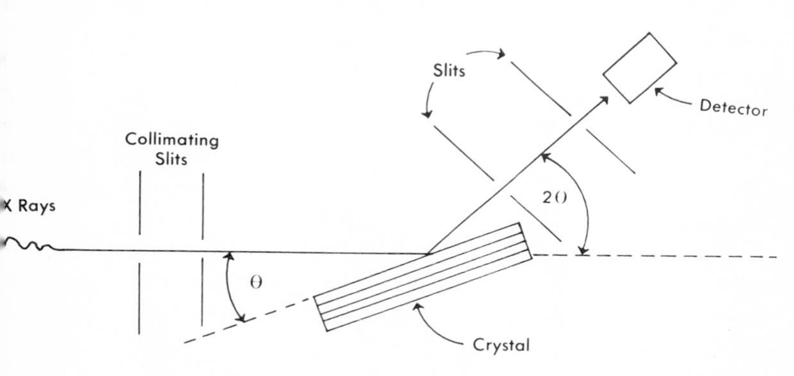

Figure 4.7. An *X*-ray spectrometer.

spectrometer utilizing Bragg's law which was derived in Chapter 2. Note that when the *X* rays are incident at an angle of Θ, the detector must be placed at an angle of 2Θ with respect to the original direction of the *X*-ray beam.

Example 4.8. Calculate the angle, 2Θ, at which the first and second order maxima should be observed for the copper and molybdenum K_α lines when incident on a calcite crystal for which the grating spacing is 3.036 Å. The wavelengths of the K_α lines for copper and molybdenum are 1.544 Å and 0.7135 Å, respectively.

Solution. For copper:
First Order:

$$\sin \Theta = \frac{\lambda}{2d} = \frac{1.544}{2 \times 3.036} = 0.2543$$

$$\Theta = 14.73^\circ \qquad 2\theta = 29.46^\circ$$

Second Order:

$$\sin \Theta = \frac{\lambda}{d} = \frac{1.544}{3.036} = 0.50856$$

$$\Theta = 30.57^\circ \qquad 2\Theta = 61.14^\circ$$

For molybdenum:

First Order:

$$\sin \Theta = \frac{0.7135}{2 \times 3.036} = 0.1175$$

$$\Theta = 6.75^\circ \qquad 2\Theta = 13.5^\circ$$

Second Order:

$$\sin \Theta = \frac{0.7135}{3.036} = 0.2350$$

$$\Theta = 13.59^\circ \qquad 2\Theta = 27.18^\circ$$

Example 4.9. It has been pointed out that the K_α line is actually a doublet, consisting of two lines close together in wavelength. For molybdenum, these wavelengths are 0.70922 Å and 0.71354 Å. Calculate the angular separation between these two wavelengths when they are measured in the first order with an *X*-ray spectrometer using a calcite crystal with grating spacing 3.036 Å.

Solution. The most direct approach is simply to calculate Θ for each wavelength and subtract.

$$\sin \Theta_1 = \frac{0.71354}{2 \times 3.036} = 0.117513 \qquad \Theta_1 = 6.7486^\circ$$

$$\sin \Theta_2 = \frac{0.70922}{2 \times 3.036} = 0.116802 \qquad \Theta_2 = 6.7075^\circ$$

$$\Delta\Theta = 0.0411^\circ\text{, or about 2.5 minutes of arc}$$

A more satisfying approach for those with some calculus is to differentiate the Bragg equation, $n\lambda = 2d \sin \Theta$, which gives $\Delta\lambda = 2d \cos \Theta \, \Delta\Theta$ so that:

$$\Delta\Theta = \frac{\Delta\lambda}{2d \cos \Theta} = \frac{0.71354 - 0.70922}{2 \times 3.036 \times 0.993}$$

since Θ is about 6.7° and $\cos \Theta = 0.993$:

$$\Delta\Theta = 0.0007165 \text{ rad} \times 57.3 \text{ deg/rad} = 0.041^\circ \text{ as above.}$$

Example 4.10. Using a Bragg type X-ray spectrometer with a NaCl crystal (grating spacing 2.82 Å), it is determined that the hort wavelength limit of the continuous X-ray spectrum comes at an angle $2\Theta = 6°$. What is the potential difference in the X-ray ube through which the electrons are accelerated?

Solution. Since $2\Theta = 6°$, $\Theta = 3°$ and $\sin\Theta = 0.052336$. Also, $\lambda = 2d \sin\Theta$, since we must use the first order, and $hc/\lambda = Ve$ so $V = hc/(2de \sin\Theta)$.

$$V = \frac{6.625 \times 10^{-34} \times 3 \times 10^{8}}{2 \times 2.82 \times 10^{-10} \times 1.6 \times 10^{-19} \times 0.052336}$$

$$= 42{,}083 \text{ volts}$$

4-12. The Moseley Experiment. In 1913, the English physicist I. G. J. Moseley (1887–1915) carefully investigated the characeristic X-ray spectra from nearly 40 elements and found that when the square root of the frequency was plotted against the tomic number of the target, a straight line was obtained. This, f course, is what one would expect on the basis of equation 4.2 nd this was the interpretation given his work at that time, proiding another significant application of the Bohr theory.

Moseley's work was particularly significant at the time, however, because it provided the best method for arranging elements n the proper order in the Periodic Table and showed the fundaental importance of the atomic number as opposed to the tomic weight. Moseley's experiments showed that several pairs f elements had been listed in incorrect order in the Periodic able on the basis of atomic weights (for example, $_{27}$Co has a reater atomic weight but a smaller atomic number than $_{28}$Ni), nd that certain Z values were missing from the Periodic Table s known at that time.

Example 4.11. A narrow beam of electrons, having been acelerated through a potential difference of 8000 volts, is incident n a thin, polycrystalline aluminum foil as in Figure 2.6. The adius of one of the rings observed on a fluorescent screen 50 cm om the foil is 2.941 cm. Calculate the spacing of the Bragg anes in the aluminum crystals which produced this ring.

Solution. The spacing d is given by the relationship $\lambda = d \sin\Theta$. We must calculate λ and $\sin\Theta$. The wavelength is given y $\lambda = \sqrt{(150/V)} = \sqrt{(150/8000)} = 0.1369$ Å. We use the classical expression for the wavelength since the kinetic energy c the electrons is a very small fraction (8/511) of the rest mass.

The radius of the ring determines 2Θ so that $\tan 2\Theta$ = 2.941/50 = 0.05882, and $2\Theta = 3.366°$ so that $\Theta = 1.683°$. Thu sin Θ = 0.02937 and:

$$d = \frac{0.137}{2 \times 0.02937} = 2.332 \text{ Å}$$

Chapter 5

The Solid State

5-1. Introduction. In this chapter, we will discuss the mechanisms by which atoms interact and are thus held together in molecules and in the various forms of the solid state. We have already calculated the spacing of atoms in NaCl and found that this spacing is of the order of the size of atoms. Therefore, we might conclude that some of the properties of solids will be determined by quantum effects because of this proximity of the atoms—despite the fact that the "system" (i.e., a typical piece of the solid) contains many, many atoms (perhaps Avogadro's number which is about 10^{23}), and is definitely not microscopic in size. We shall be interested in why certain materials are good conductors of electricity and others are not—and in the interesting properties of semiconductors which are inbetween. We shall be primarily interested in crystalline materials in which the atoms are arranged in a regular repetitive array.

5-2. The Ionic Bond. One type of bond which holds atoms together in molecules is called the ionic bond and is explained very simply in terms of Coulomb's law. When elements in the first column of the Periodic Table (Li, Na, K, Rb, and Cs) combine with those in the seventh column (F, Cl, Br, and I), the result is that an element such as chlorine takes an electron from sodium, leaving sodium as a positive ion and chlorine as a negative ion. These ions then attract each other and a sodium chloride "molecule" is formed. (To limit our discussion to a single molecule, we are talking about sodium, chlorine, and sodium chloride in the gaseous state.) It takes 5.14 ev to remove the valence electron from sodium, while adding an extra electron to chlorine releases .82 ev. This released energy is called the *electron affinity* and is a measure of the tendency of an atom to attract an electron in excess of its normal number. Atoms with large electron affinities, such as chlorine, tend to pick up electrons from atoms, such as sodium, with low electron affinities. As indicated in Table 4-2, page 78, chlorine is one electron short of the closed shell configuration of the noble gas argon, while sodium has one more electron than the closed shell neon configuration. Thus, elements with a few electrons outside a closed shell tend to give them up to an element with a nearly closed shell with each approaching the more stable noble gas configuration.

Figure 5.1 gives the potential energy of an ionic molecule such as NaCl as a function of the separation between the atoms. When

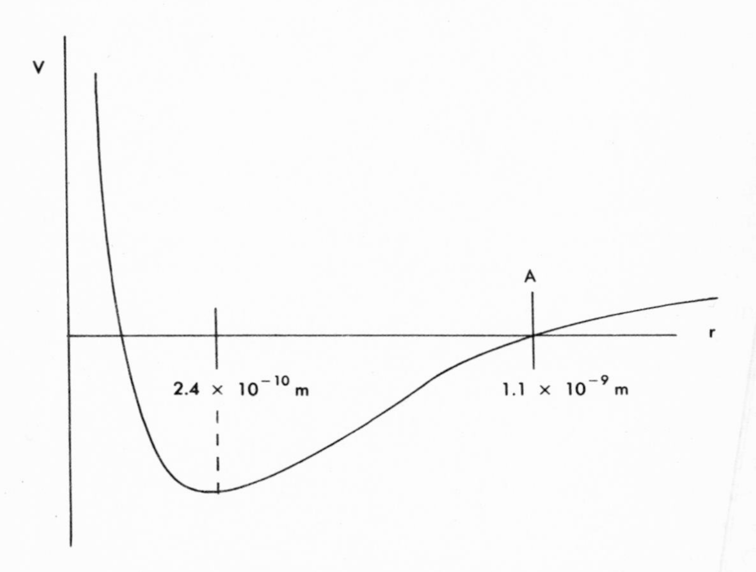

Figure 5.1. Potential energy for the NaCl molecule (only roughly to scale).

they are very far apart, we have seen that 5.14 − 3.82 = 1.32 ev are required to form the positive and negative ions from the neutral atoms. As they are brought closer together, energy is released and, at a separation of 1.1×10^{-9} meters, the total energy of the system is zero since the electrostatic potential energy is −1.32 ev.

Example 5.1. Show that the electrostatic potential energy between charges +e and −e is −1.32 ev at a separation of about 11×10^{-10} m.

Solution. We use $V = \left(\frac{1}{4\pi\epsilon_0}\right)\left(\frac{e^2}{r}\right)$ and recall that 1 ev = 1.6×10^{-19} joules.

$$r = \frac{e^2 \times 9 \times 10^9}{1.32 \times 1.6 \times 10^{-19}}$$

And as $1/4\pi\epsilon_0 = 9 \times 10^{+9}$ newton m^2/coul^2:

$$r = \frac{1.6 \times 10^{-19} \times 9 \times 10^9}{1.32} = 11 \times 10^{-10}\,\text{m}$$

If the ions are brought closer together than 11×10^{-10} m, the potential continues to decrease, indicating that they continue to attract one another until they get close enough that the electron clouds of the two atoms start to overlap appreciably. At this point, a repulsive force begins which is due in part to the repulsion of the two positive nuclei, but a more important effect arises because of the Pauli exclusion principle. This is a strictly quantum mechanical effect and is caused by the fact that, as the electron systems for the two atoms begin to overlap appreciably, we begin to approach a single quantum mechanical system in which more than one electron would have to have the same set of four quantum numbers. To avoid this, some electrons would have to go into quantum states of higher energy than they would normally occupy. This requires energy so that to move the two systems together requires that work be done, and therefore a large repulsive force. At the minimum of the potential energy curve of Figure 5.1, the ions in the NaCl molecule experience no force, but work must be done to move them apart and to push them closer together and thus they are "bound" together. This minimum occurs at about 2.4×10^{-10} m for a single molecule, which is somewhat less than the distance between atoms for solid NaCl which we calculated in Chapter 2.

Example 5.2. Estimate the energy required to pull apart a NaCl molecule into neutral Na and Cl atoms.

Solution. The coulomb energy at the minimum of the potential energy curve of Figure 5.1 is 6 ev, which we calculate as follows:

$$V = \left(\frac{1}{4\pi\epsilon_0}\right)\left(\frac{q^2}{r}\right) = \frac{9 \times 10^9 \times (1.6 \times 10^{-19})^2}{2.4 \times 10^{-10} \times 1.6 \times 10^{-19}}$$

$$= -6\ \text{ev}$$

If the ions are separated to a distance of 11×10^{-10} m (point A, Figure 5.1) where the electrostatic energy is -1.3 ev, we must do work amounting to about 4.7 ev. Since the difference between the ionization potential for sodium and the electron affinity for chlorine was needed to make the Na^+ and Cl^- ions when the molecule was formed, we should get this much energy back when

we replace the electron that the chlorine took from the sodium
Thus, we must subtract another 1.3 ev (5.1 – 3.8), obtainin
4.7 – 1.3 = 3.4 ev.

The ionic bonds holding together molecules such as NaCl, KB
and ZnO also bind the ions in crystalline solids as indicated i
Figure 2.4 for NaCl. Each ion tends to be surrounded by as man
ions of opposite charge as possible, and ions of the same sign ar
as far from each other as possible. Ionic bonds are quite stron
and, therefore, ionic crystals are strong and have high meltin
points.

5-3. The Covalent Bond. Another type of bond that can exis
between atoms is the covalent bond which results from the shar
ing of one or more pairs of electrons. The hydrogen molecule i
an example of a strictly covalent bond. We picture the two elec
trons as moving about the positive nuclei as indicated in Figur
5.2 realizing, of course, that representing the electrons as particle

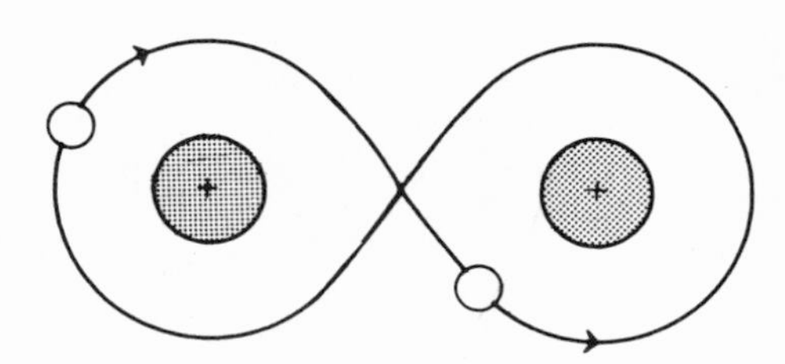

Figure 5.2. The covalent bond for the H_2 molecule.

is a gross oversimplification. It turns out that the two electrons
on the average, spend more time between the nuclei than they d
on the ends, with the result that there is an appreciable averag
negative charge between these nuclei which is sufficient to attrac
the positive nuclei and more than offset the repulsion betwee
the like positive charges. In a certain sense, we see that th
covalent bond is also electrostatic in nature. As in the case o
the ionic bond, there will be a certain equilibrium separation a
which the repulsion of the nuclei just balances the covalent at
tractive force. For the hydrogen molecule, this occurs at a separa
tion of about 0.74×10^{-10} m. Again, it is interesting to note tha
the covalent bond is limited to two electrons with oppositel
directed spins since the Pauli exclusion principle permits only tw
in the same energy state.

Carbon atoms have four electrons in their outer shell and tend to form four covalent bonds with various other atoms. Well over a million compounds involving carbon covalently bonded with other atoms such as hydrogen, oxygen, nitrogen, and phosphorus can be formed. Complex cyclic chain structures can be formed containing many thousands of atoms. Organic chemistry is concerned with the study of such compounds. Some simple examples are shown in Figure 5.3.

Methane   Benzene (C_6H_6)   Napthalene ($C_{10}H_8$)

'igure 5.3. Some simple organic compounds. The double lines indicate double bonds, each bond involving a pair of electrons.

5-4. Van der Waals Forces. An interesting quantum mechan- cal effect provides weak bonding for certain molecules (and cer- ain atoms) which do not have electronic structures of a type that an lead to ionic or covalent binding. It is easy to understand how polar molecule such as HCl might attract another such mole- ule. By polar we mean that the electric charge distribution in the molecule is not symmetric and is arranged with an excess of ositive charge at one end and an excess of negative charge at ne other. It is natural to expect that the positive portion of one molecule will attract the negative portion of the other and even o expect that a polar molecule can induce a polarization in a earby nonpolar molecule and thereby attract it. Actually, the harge distribution in all atoms and molecules varies as a function f the time so that even if the average distribution is spherically ymmetric, the actual distribution at a particular instant may not e. These variations can influence the variations of the charge dis- ibution in a nearby molecule or atom so that the two change nore or less together so as to result in a slight average attraction milar to that between two polar molecules.

Van der Waals bonds are much weaker than ionic or covalent onds. Therefore, the solids into which rare gases such as argon and neon crystallize are weak and can only exist at low temperatures (such that thermal energies are small). The melting point of argon is 87° K, while that of krypton, another noble gas, is 120° K.

Crystals whose bonding is due almost entirely to ionic or metallic (see Section 5-5) forces also experience Van der Waals forces but these rarely amount to more than a few percent of the binding energy of the crystal (more properly called the cohesive energy).

5-5. The Metallic Bond. Materials which we describe as metals are characterized by the fact that they have only a few electrons in the outermost shell at least one of which can be easily removed. We picture a metal then as a lattice of positive ions through which moves a "gas" of electrons which have left these ions. An attractive force arises from the interaction of the negative cloud of electrons which is spread out through the metal and the positive ions, producing a situation which is very roughly similar to the covalent bond except that the electrons are not associated with any one atom but move about through the crystal.

5-6. The Band Theory of Solids. Atoms in solid crystals are so close to each other that the outer or valence electrons of the various atoms interact and, as we have stated in the previous section, form a single system or "gas" of electrons. The inner shells are relatively unaffected. The energy levels for the valence electrons in the isolated individual atoms are, of course, quantized but, as the atoms are brought closer and closer together, these levels are spread out into bands of closely-spaced allowed energy states. These energy states are so numerous as to be virtually continuous. This effect is illustrated in Figure 5.4.

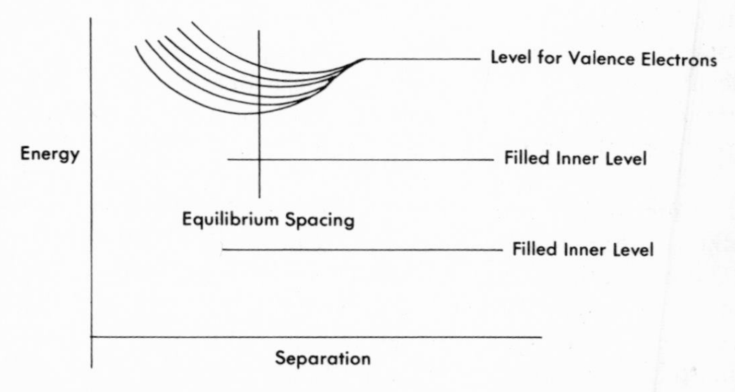

Figure 5.4. Splitting of an energy level for the valence electrons as a function of the separation of atoms in a crystal.

In general, we would expect that the band of levels in the solid vould be lowered so as to provide the bonding mentioned in Section 5.5, that levels should rise at small atomic spacing because of the exclusion principle, and that the equilibrium spacing should correspond to the low energy point or the minimum in the band of energies as indicated in Figure 5.4. The position in energy of an allowed band in a solid corresponds roughly to the position of a particular energy level in the atom. The properties of crystalline solids depend very strongly on the relative spacing of these bands and the number of electrons which normally occupy the bands.

Figure 5.5 illustrates, in a general way, the relationship for conductors, insulators, and semiconductors between the highest

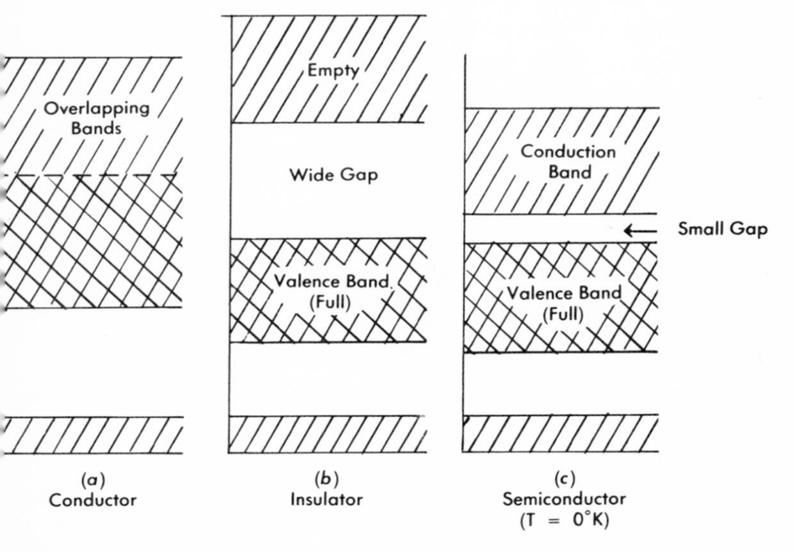

Figure 5.5. Energy level diagram showing relative positions of energy bands for (*a*) conductors, (*b*) insulators, and (*c*) semiconductors.

band for the valence electrons and what is called the conduction band. In the conduction band, electrons are frec to move through the crystal under the influence of an applied electric field. Non-conductors of electricity are characterized by a relatively wide energy gap between the filled valence band and the conduction band, while good conductors have overlapping partially-filled conduction and valence bands and semiconductors have a relatively narrow gap between these bands.

For example, in diamond, a good insulator, the band gap is about 7 ev, which means that an electric current can flow only if an additional 7 ev is provided to electrons in the valence band. This is possible using energetic photons or other energetic particles; but at normal temperatures, thermal energies are insufficient to lift electrons across this gap. In a semiconductor such as germanium, however, the gap is only 0.7 ev. It should be clear that only if a band is partially filled is conduction possible, since in a filled band there are no states available to electrons and they are unable to move. One might think that, by applying a potential difference of considerably more than 7 volts across an insulator such as diamond, electrons should be raised across the band gap. However, electrons not in the conduction band can go only very short distances before colliding with another atom or electron and losing their kinetic energy. Thus, extremely high voltages (about 600 million volts per meter) are required to raise electrons into the conduction band if the average distance an electron can move before experiencing a collision is taken to be about 10^{-8} m.

5-7. The *p-n* Junction and Semiconductor Diodes and Transistors. The properties of semiconductors, primarily silicon and germanium, can be utilized in an interesting way to construct what are known as solid state diodes and transistors which are such an important part of modern technology. These devices are based on what is known as the *p-n* junction which we now describe in terms of the electron configuration of silicon or germanium atoms and the resulting band structure of the covalent germanium and silicon crystals.

Both silicon and germanium have four valence electrons which form four covalent bonds with their neighbors in the solid crystal. Silicon has a band gap of 1.1 ev while the gap for germanium is 0.7 ev. Thus, they are not particularly good conductors at room temperature. The conductivity can be increased appreciably in a very interesting way by adding the right kind of impurities to these semiconductors. For example, if we were to add a small amount of arsenic to germanium and then allow a crystal to form, arsenic atoms, being about the same size as germanium atoms, could easily replace germanium atoms in the crystal—with the important difference that arsenic has a nuclear charge one unit larger than germanium and thus has one more valence electron so that after the four covalent bonds are formed we have an electron left over which is very loosely bound to the arsenic atom. Similarly, if we had added indium (with three valence electrons) instead of arsenic, we would be one electron short in forming the four covalent bonds in the crystal and such a deficiency is called a "hole."

Figure 5.6 indicates that the extra electron in the case of the arsenic impurity lies very close to the conduction band. Therefore,

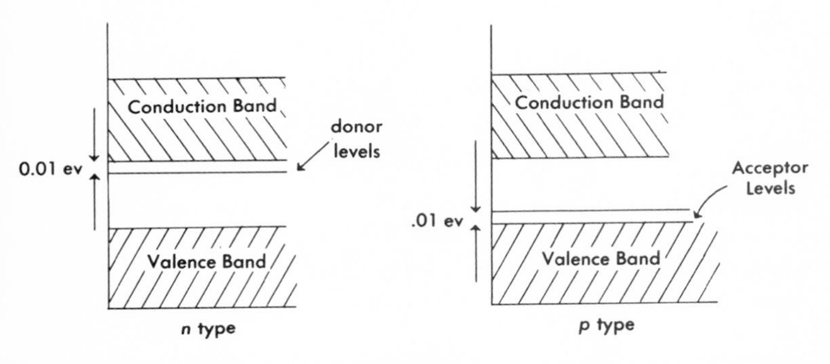

Figure 5.6. Location of donor and acceptor levels resulting from impurities in *n* type and *p* type semiconductors.

it is very easily lifted up to the conduction band providing a carrier and increasing the conductivity. A semiconductor with such an impurity is called an *n* type semiconductor.

In the case of the indium impurity, the "holes" represent a missing electron and, therefore, it is possible for an electron from a neighboring atom to move into this vacancy, providing a mechanism for current flow. If this is the case, the hole has moved to another atom, and the situation is just as though positive electrons were moving through the crystal, although actually electrons are moving from hole to hole. When impurities are added so that holes provide the mechanism for conduction, the semiconductor is called *p* type, and the levels into which electrons may move are called acceptor levels as indicated in Figure 5.6.

At room temperature, thermal energy is available to provide some carriers of both types, and these are called intrinsic holes and electrons. In *n* type and *p* type semiconductors, it is assumed that the impurity carriers are much more numerous than the intrinsic carriers. Because of the small band gap, however, the number of intrinsic carriers is extremely dependent on the temperature. In the case of germanium, if the temperature is raised to 100° C, the number of intrinsic carriers becomes much larger than the number of impurity carriers no matter what the impurity. For this reason, great care must be taken to limit the range of operating temperatures for semiconductor devices.

Most interesting applications of semiconductors involve the *p-n* junction which is formed by joining pieces of *n* and *p* type semiconductors as a single crystal. Such junctions must be grown or made by diffusion, and cannot be produced by simply mechanically joining bars of *n* and *p* type materials. When a junction is "formed," the free electrons on one side and the holes on the other will diffuse into the other material with the effect illustrated in Figure 5.7. On the *p* type side, the electrons diffusing across will

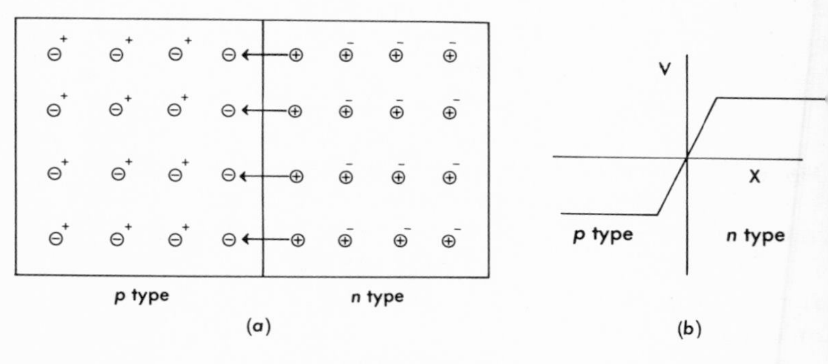

Figure 5.7. A *p-n* junction.

neutralize holes, leaving fixed negative charges on those impurity atoms near the junction. These electrons leaving the *n* side must leave fixed positive charges on the impurity atoms from which they came. Thus, a potential difference is set up at the junction with polarity such as to oppose this diffusive motion. An equilibrium situation results which depends on the number of impurity atoms in the *n* and *p* type materials.

Such a junction is said to be reverse biased if an external potential difference is applied as shown in Figure 5.8*a* such that the external potential difference adds to the potential difference at the junction. The junction is forward biased if the external voltage is such as to reduce the natural junction voltage. Figure 5.9 shows the current-voltage characteristics of such a *p-n* junction. It is seen that when forward biased small applied voltages

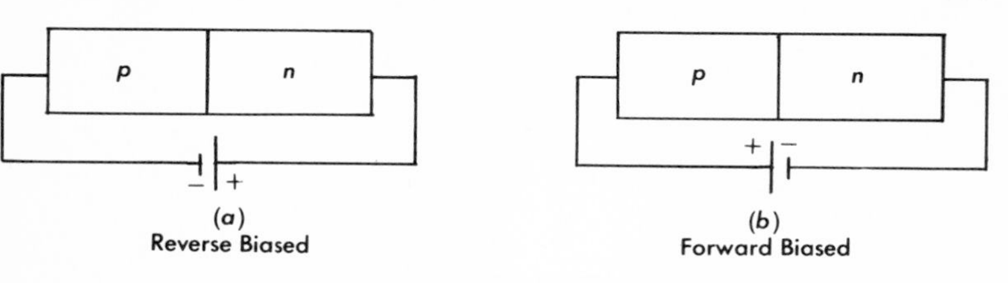

Figure 5.8. Reverse biased (*a*) and forward biased (*b*) *p-n* junctions.

result in large currents, but when reverse biased virtually no current flows until such large voltages are applied that the junction breaks down. This is easily understood in terms of Figure 5.7. When reverse biased, the carriers on either side are pulled away from the junction so that at the junction no carriers are available except for intrinsic carriers. When forward biased, the carriers on either side of the junction move toward the junction, carriers are available, and large currents are possible.

A single *p-n* junction is called a diode since it behaves much like a vacuum tube diode in that it has practically zero resistance in one direction and almost infinite resistance in the other.

A simple and practical application of the diode is the conversion of an alternating voltage to a pulsating voltage of one

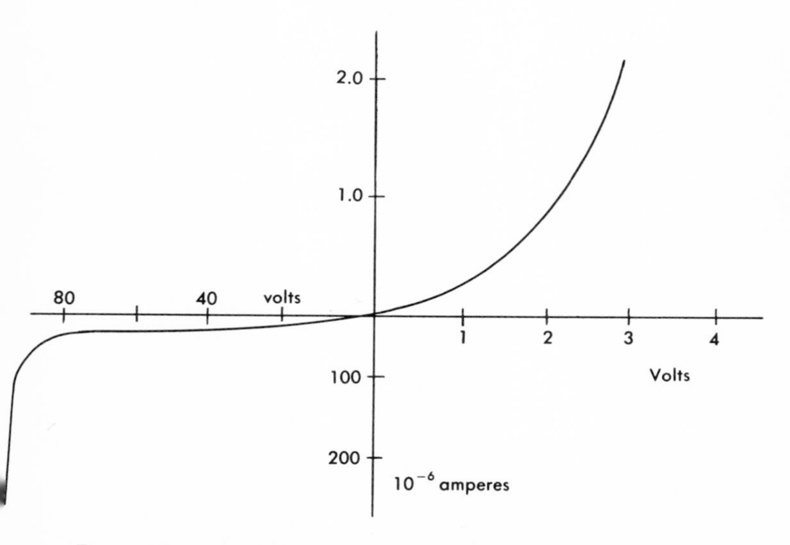

Figure 5.9. The current-voltage characteristics of a *p-n* junction.

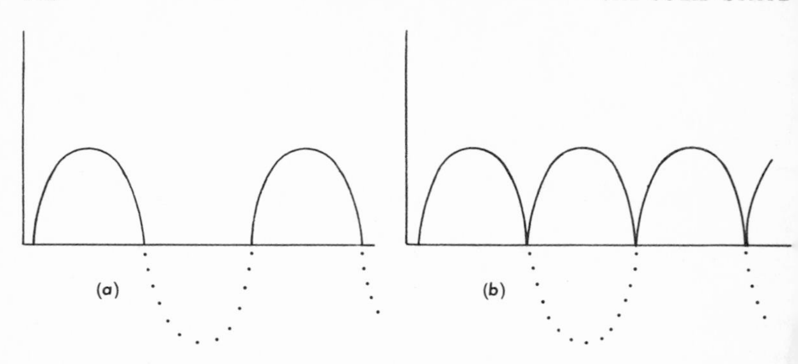

Figure 5.10. Half wave (*a*) and full wave (*b*) rectification.

polarity. Three possible arrangements are shown in Figure 5.11 using one, two, and four diodes. This process of rectification is perhaps best understood by reference to Figure 5.10*a*, in which the solid line represents the applied alternating voltage in the circuit of Figure 5.11*a* and the dotted line gives the voltage across the resistor R. Note that for one polarity the diode conducts, while for the other it does not, thus cutting off the negative half of the A.C. voltage and providing a pulsating D.C. voltage.

One can obtain rectification of both halves of the A.C. cycle using the circuit of Figure 5.11*b* at the expense of a factor of two in the voltage. In Figure 5.11 the A.C. input is assumed to be the secondary of a transformer and in Figure 5.11*b* the resistor R is connected at its midpoint. The factor of two decrease can be eliminated by using four diodes as in Figure 5.11*c*. As the polarity of the alternating voltage changes, current flows through a different pair of diodes, but the current through the resistor R is always in the same sense. The arrows indicate the situation for one polarity.

Transistors are devices which utilize two *p-n* junctions as indi-

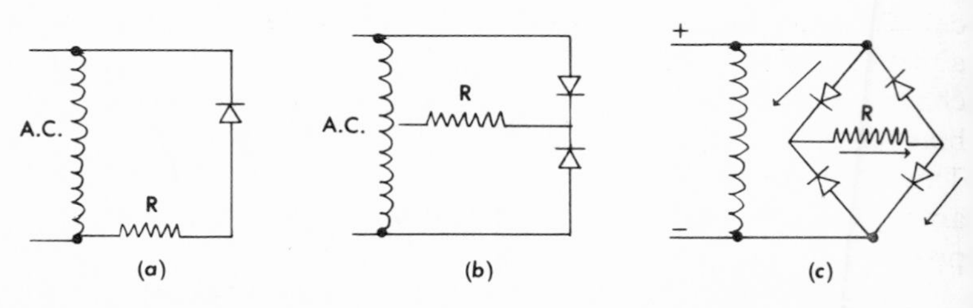

Figure 5.11. Circuits for (*a*) half wave, (*b*) full wave, and (*c*) bridge rectification.

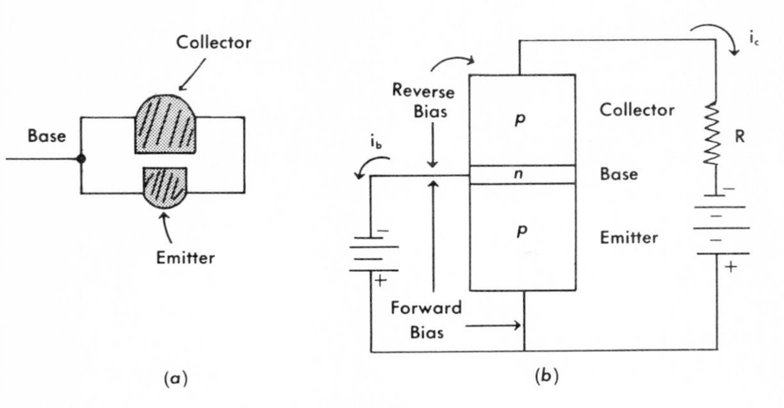

Figure 5.12. The transistor.

cated schematically in Figure 5.12. They may be of the *p-n-p* type in which a narrow *n* type region is included between two *p* type regions or just the reverse, which is called the *n-p-n* type. Figure 5.12*b* indicates one manner in which a transistor might be used electrically to act as an amplifier. We have *p-n* junctions between both the base and emitter and the base and collector. These portions of the transistor are indicated in Figure 5.12, the names having evolved during the development of the transistor. We refer to Figure 5.12*b* to discuss the basic operation of the transistor. Note that the junction between base and collector is reverse biased and that the emitter-base junction is forward biased. The impurities are added to the emitter and base so that most of the current that flows when the emitter-base junction is forward biased consists of holes from the base. Since the collector-base junction is reverse biased, any holes in the base will be attracted to the collector. If the base is very thin, most of the holes injected from the emitter will go right across to the collector and the resulting current in the collector circuit will flow through the resistance R. The current in the connection to the base should be very small since almost all of the holes go to the collector. Therefore, very small changes in the base current, reflecting changes in the emitter-base bias, result in large changes in current in the collector circuit. Thus, the transistor can control the flow of a relatively large amount of power with the expenditure of a small amount of power.

Transistors and solid state diodes have permitted tremendous strides in the development of modern communications and computing equipment because of their small size, low cost, and great reliability. In general, silicon is best for many uses because of its wider band gap and resultant insensitivity to temperature changes as compared to germanium. However, the latter material is, in general, better for high frequency and fast switching applications because holes and electrons travel more rapidly in germanium.

Small chips of silicon having surface areas of about one or two square centimeters can be treated chemically and optically to produce perhaps 100 diodes or transistors on this area. Such miniaturization is essential, particularly in the development of faster computers since transistors can be made which can switch electrical signals in about 10^{-9} seconds. If logical operations are to be made in times of a few nanoseconds, those portions of the computer which must communicate with each other electrically must be physically close enough so that operations are not slowed down by the fact that signals traveling at the ultimate speed limit, the velocity of light, can only go about one foot in one nanosecond (10^{-9} seconds).

5-8. The Specific Heat of Solids. When we consider a solid composed of a large number N of atoms arranged in a regular manner in some sort of crystal lattice, it should be clear that, if disturbed, each atom can vibrate about its equilibrium position and in so doing, disturb its neighbors and pass on the disturbance as a wave in the crystal. Classically, as is done in the kinetic theory of gases, we would associate an energy $\frac{1}{2}k\text{T}$ with the kinetic energy of vibration and $\frac{1}{2}k\text{T}$ with the potential energy for each atom. Since there are three "degrees of freedom," i.e., three mutually perpendicular directions along which vibrations may occur, we associate an energy $k\text{T}$, where k is the Boltzmann constant, with each. This gives a total thermal energy of $3k\text{T}$ per atom or $3N_0k\text{T}$ per mole, where N_0 is Avogadro's number.

Recalling that R, the universal gas constant, is equal to N_0k, the thermal energy of one mole of a crystalline solid should be $3R\text{T}$. Since the specific heat (the amount of heat required to increase the temperature of one mole by one centigrade degree) at constant volume is given by $C_v = \Delta\text{V}/\Delta\text{T}$, we get $C_v = 3R$. R is approximately equal to 2 cal/mole-°C, so that classically we would expect a molar heat capacity of about 6 cal/mole-°C. This is known as the *law of Dulong and Petit* and gives fairly good

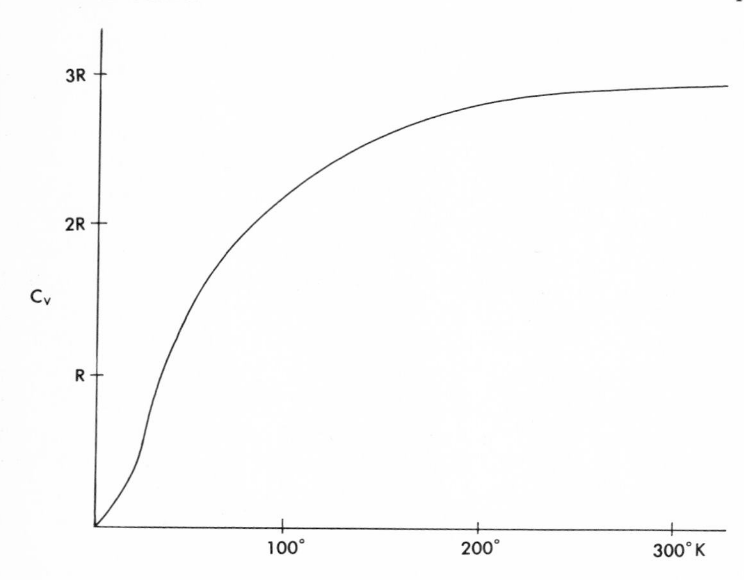

Figure 5.13. The behavior of the specific heat of a typical conductor as a function of temperature.

agreement at temperatures near 300° K, but, as indicated in Figure 5.13, the low temperature measurements do not agree. (In fact, C_v approaches zero at absolute zero.)

This behavior of C_v as a function of temperature has been successfully predicted by a theory, developed by P. Debye in 1912, which requires that the vibrational motion of the atoms in the crystal be quantized so that atoms can lose or gain vibrational energy only in discrete amounts which are propagated through the crystal as bundles of acoustic energy known as phonons (see page 68).

We have pointed out in Section 5-5 in our discussion of the metallic bond that conductors have, on the average, about one free electron per atom which can move freely through the crystal as part of an electron gas which actually is one quantum system. One would expect classically that if this is the case, the average kinetic energy of these electrons should increase as heat is added to a metal with the result that the specific heat of the metal should be larger. If the electrons behave like an ideal gas, their average kinetic energy should be $\frac{3}{2}kT$ and should make a contribution to the specific heat of $\frac{3}{2}R$ so that the Dulong-Petit value should be $\frac{9}{2}R$ instead of 3R.

Clearly the electrons do not contribute substantially to the specific heat of a metal since metals have specific heat values close to the Dulong-Petit value, and the Debye theory, which neglects the electrons, gives good agreement for both conductors and nonconductors. The explanation for this apparent paradox is provided by the Pauli exclusion principle which must be applied to these free electrons in conductors if they are to be considered as all belonging to the same quantum system. No two electrons may have the same set of quantum numbers and, therefore, when the lower energy states are filled, the other electrons must go to higher energy states. The distribution of electron energies in a metal is shown in Figure 5.14 for various temperatures. It is seen that at absolute zero, contrary to the situation in an ideal gas, the electrons do not all have zero energy, but exist in states of increasing energy up to a maximum known as the Fermi energy. For copper the Fermi energy is about 7 ev, while for sodium it is 3.1 ev. These values are very large compared to the classical ideal gas value of $\frac{3}{2}kT \approx 0.04$ ev for the average thermal kinetic energy of such an electron. For this reason, only electrons near the Fermi energy can acquire thermal energy from the crystal since such energy changes must be of the order of kT. For most of the electrons, no states are available within an energy range of kT, and hence most of the electrons do not interact thermally with the crystal. For this reason they do not contribute to the specific heat.

At temperatures above absolute zero, one would expect that the distribution would be rounded near E_f over a range of about kT and that the conduction electrons would make a small but increasing contribution to the specific heat as the temperature is increased. The Fermi energy in copper corresponds to a temperature of about 80,000° K, indicating the tremendous effect of the uncertainty principle in this situation.

5-9. Electrical Conductivity. For metals it should be clear that the large electrical conductivities which are experimentally observed must be due to the "gas" of free electrons which we have discussed in the preceding sections. Since these electrons obey the exclusion principle and have an energy distribution, known as the Fermi distribution, as shown in Figure 5.14, only those electrons

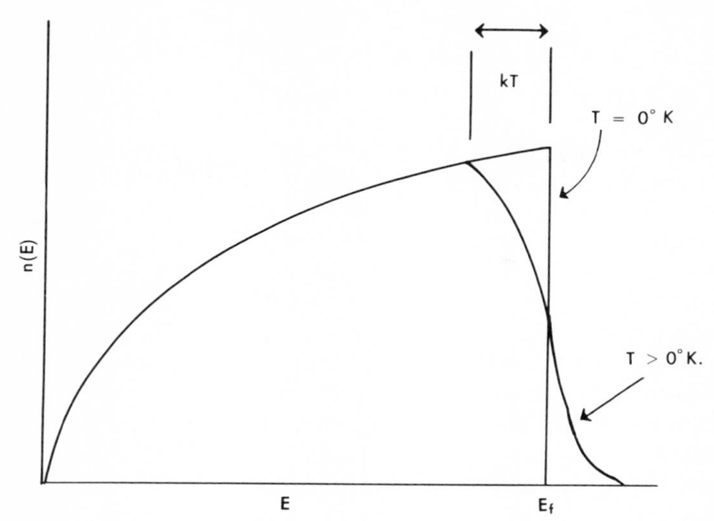

Figure 5.14. The distribution of electron energies in a conductor at absolute zero and some higher temperature.

at the high energy side of the distribution can contribute to electrical conduction. This is because they are the only electrons for which empty energy states are available in view of the small changes in energy which the electrons experience when an electric field is applied across a conductor. It is a well-known fact that the resistance of a particular conductor is proportional to the absolute temperature. Clearly, the temperature associated with the electrons at the Fermi energy is so large that the small changes which affect the resistance experimentally should have a negligible effect. For copper, for example, a change of 100° out of 80,000° represents a change of 0.125%, rather than a 30% change when compared with room temperature (300°K).

The explanation for the temperature dependence of resistance is provided by the wave nature of matter and has to do with the fact that waves are propagated by a regular periodic lattice in a crystal with no scattering and, thus, zero resistance if the crystal is perfect. This may be understood in a very qualitative manner by considering the Bragg equation, $n\lambda = 2d \sin \Theta$, for electrons in the crystal. The wavelength of the electrons, even those at the Fermi energy, is larger than the crystal spacing so that $\lambda/2d$ is greater than one. Thus, even for $n = 1$, no scattering is possible if the crystal is perfect. However, if irregularities such as atoms of another element are present in the crystal or if atoms are missing in the lattice, the regularity of the crystal is spoiled and scattering may take place which we "explain" by associating d with the distance between imperfections which can be such that 2d is greater than λ. Furthermore, unless the temperature is at absolute zero, the atoms are vibrating about their equilibrium positions and this, too, spoils the regularity of the crystal.

The effect of this vibration on the resistance may be measured by the increased cross-sectional area presented. This area is proportional to the square of the amplitude of vibration. The amplitude is proportional to the square root of the energy of vibration and the energy is proportional to the temperature so that the cross-sectional area, and thus the resistance, is proportional to the temperature in agreement with experiment. Thus, the temperature dependence of electrical resistance is another microscopic phenomenon which must rely on quantum mechanics for a proper explanation.

Problems

Chapter 1.

1. Two parallel plates 10 cm long are separated by 4 cm. An electron moving with a velocity of 2×10^7 m/sec moves parallel to the plates as it enters the region between the plates halfway between them. What potential difference must be applied between the plates so that as the electron leaves the region between the plates it just misses one of the plates?

   Ans. 364 volts.
2. A charged oil drop of mass 10^{-13} kg is just balanced in the electric field between two parallel plates. What electric field must be supplied between the two plates to accomplish this if the drop has a charge of 10 electronic charges?

   Ans. 6.13×10^5 volts per meter.
3. Calculate the distance of closest approach for an alpha particle of mass 6.68×10^{-27} kg and velocity 2×10^7 m/sec incident on an aluminum nucleus ($Z = 13$).

   Ans. 4.5×10^{-15} m.
4. Neglecting the buoyant force of the air, show that the electric field, E, necessary to raise an oil drop of mass, m, and charge, q, with a speed three times the free fall speed when $E = 0$ is $E = 4\,mg/q$.
5. Calculate the energy in electron volts of an alpha particle accelerated through a potential difference of 1000 volts.

   Ans. 2000 ev.
6. Calculate the radius of curvature of 10,000 ev protons in a magnetic field of 1.371×10^{-3} webers/m^2. *Ans.* 10.5 m.
7. Calculate the factor by which the half life of a π-meson is increased if its velocity is $0.6c$. *Ans.* 1.25.
8. Calculate the factor by which the mass of an electron has increased when it has a kinetic energy of 1000 Mev. *Ans.* 1958.
9. Calculate $\beta = v/c$ for an electron, proton, and alpha particle, each of which has an energy of 100 Mev.

   Ans. 0.999987, 0.4282, 0.2271 (The rest mass of alpha particle is 3727 Mev.)
10. Calculate the momentum, in units of Mev/c, of a 50 Mev electron, proton, and alpha particle.

    Ans. 50.5 mev/c, 310.3 mev/c, 612.5 mev/c.

11. The mass of the neutron is 1.008665 amu and the mass of the hydrogen atom is 1.007825 amu. Calculate the binding energy of the carbon nucleus. (1 amu = 931.48 Mev.)
Ans. 0.09894 amu = 92 Mev.
12. Silver has an atomic weight of 107.87. If the isotopes ^{107}Ag and ^{109}Ag have masses of 106.905 and 108.905, what is the percent abundance of each in ordinary silver?
Ans. 51.75% and 48.25%.

Chapter 2.

1. Calculate the work function in electron volts for a particular substance if the maximum kinetic energy photoelectrons are found to have an energy of 0.76 ev when the surface is illuminated by light of wavelength 3300 Å. *Ans.* 3 ev.
2. A photon of energy 5 Mev is Compton scattered at an angle of 37°. If its momentum after scattering is 1.675 Mev/c, calculate the angle at which the Compton electron is ejected with respect to the incident photon direction. (*Hint:* Recall that the energy and momentum of a photon are equal if the energy is expressed in Mev and the momentum in Mev/c.) *Ans.* 15.35°.
3. Calculate the wavelength of a photon Compton scattered through an angle of 30° if the original wavelength was 0.02 Å.
Ans. 0.0233 Å.
4. A 10 Mev gamma ray produces a positron-electron pair. If the positron has twice the energy of the electron, calculate the energy of the electron. *Ans.* 3 Mev.
5. Calculate the voltage required to accelerate an electron such that its de Broglie wavelength is 0.707 Å. *Ans.* 300 volts.
6. In an electron diffraction experiment such as illustrated in Figure 2.6, the screen is located 50 cm from a thin polycrystalline aluminum foil. Electrons of energy 8000 ev are incident on the foil, and the radius of one of the rings is measured to be 3.4 cm. What spacing in the aluminum crystal corresponds to this ring? *Ans.* 2.02 Å.
7. Consider an experiment such as illustrated in Figure 2.7. If the width of the slit is 10^{-6} m and the wavelength of the incident electrons is 1 Å, use the uncertainty principle to calculate the sine of the angle Θ corresponding to the first minimum.
Ans. 10^{-4}.

Chapter 3.

1. Calculate the energy of the $n = 4$ energy state of the hydrogen atom in electron volts relative to the $n = 1$ or ground state. *Ans.* 12.73 ev.
2. Calculate the wavelength in Angstrom units of the shortest wavelength in the hydrogen spectrum, i.e., the series limit for the Lyman spectrum. *Ans.* 921 Å.
3. Calculate the longest wavelength in the Lyman series, i.e., the series for transitions ending at the ground state. *Ans.* 1215 Å.
4. Show that the radius of the first Bohr orbit is 0.529 Å.
5. Show that the velocity of the electron in the first Bohr orbit is about 1/137 the velocity of light.
6. Calculate the reduced mass for a triply ionized beryllium atom. *Ans.* 0.99994 m_e.
7. Calculate the radius of the first Bohr orbit for triply ionized beryllium. *Ans.* 0.132 Å.
8. Show that the energy difference of 4.86 ev obtained in the Franck-Hertz experiment for the first ionization potential of mercury corresponds to a wavelength of about 2537 Å in the mercury spectrum.
9. A photon of wavelength 4000 Å is absorbed by an atom. The atomic system is raised from one energy state to a higher energy state. What is the energy difference between these two states? *Ans.* 3.1 ev.
0. Calculate the energy width of an excited state in an atom if it is known that the lifetime of the state is about 5×10^{-8} sec. *Ans.* 8×10^{-8} ev.

Chapter 4.

. Consider an electron with an orbital angular momentum quantum number $l = 4$. Calculate the magnitude of the angular momentum in units of $\hbar$ and determine the possible values of m_l. *Ans.* $\sqrt{20}\hbar$; $m_l = 4, 3, 2, 1, 0, -1, -2, -3, -4$.

. Using spectroscopic notation, give the ground state electronic configuration for aluminum. *Ans.* $1s^2, 2s^2, 2p^6, 3s^2, 3p^1$.

. Calculate the short wavelength limit of the continuous X-ray spectrum produced by 80,000 volt electrons incident on a target. *Ans.* 0.155 Å.

. Calculate the wavelength of the K_α line in the characteristic

X-ray spectrum of copper, and compare it with the accepted value of 1.537 Å. (For copper, Z = 29.) *Ans.* 1.550 Å.

5. Use the Bohr model to estimate the wavelength of the K_α line for silver (Z = 47), and compare it with the accepted value of 0.558 Å. *Ans.* 0.574 Å.
6. When an L electron falls into a vacant K shell position in a lead atom, an X ray of 75,030 ev is given off. If the binding energy of this L electron is 13,070 ev, calculate the binding energy of the K electron. What minimum potential difference is required for an X-ray tube with a lead target in order to produce the K X ray mentioned above?
Ans. 88,100 ev and 88,100 volts.
7. Calculate the wavelength of the characteristic X ray mentioned in problem 6. What is the wavelength of the most energetic X ray in the continuous spectrum produced when electrons are accelerated through a potential difference of 88,100 volts in an X-ray tube with a lead target? Would this wavelength be different if the target were made of copper?
Ans. 0.166 Å, 0.141 Å, no.
8. A Bragg type X-ray spectrometer, as shown in Figure 4.4, is used to measure the wavelength of the K_α line from a molybdenum target. A calcite crystal is used in which the grating spacing is 3.0356 Å. If the first order K_α reflection is observed at an angle 2θ of 13.4 degrees, what is the wavelength of the K_α line from molybdenum? *Ans.* 0.709 Å.
9. Using the results of problem 8, calculate the angle at which the spectrometer should be set to observe the first order K_α peak if the crystal is rock salt instead of calcite. (Rock salt has a grating spacing of 2.82 Å.) *Ans.* 14.4°

Appendix 1

A. Useful Conversion Factors

1 newton = 0.244 lb
1 meter = 100 cm = 39.37 in = 3.28 ft
1 inch = 2.54 cm
1 mile = 5280 ft = 1.61 km
1 radian = 57.296 degrees
1 gram calorie = 4.186 joules
1 kilogram calorie = 3100 ft lbs = 4186 joules
1 horsepower = 746 watts
1 electron volt = 1.602×10^{-19} joules
Ratio of proton rest mass to electron rest mass = 1836.12
1 weber/m^2 = 10^4 Gauss
1 year = 3.156×10^7 sec = 5.259×10^5 min = 8.766×10^3 hrs
1 year = 365.2 days
1 Angstrom unit = 10^{-10} m
1 atomic mass unit = 1.66×10^{-27} kg = 931.48 Mev

B. Fundamental Constants

Velocity of light in vacuum	c	= 2.997925 m/sec
Planck's constant	h	= 6.625×10^{-34} joule sec
Avogadro's Number	N_0	= 6.02252×10^{23} atoms/g at wt
Molar Gas constant	R	= 8.3143 joules/mole-K°
Boltzmann constant	k	= 1.3805×10^{-23} joules/K°
Charge on electron	e	= 1.6021×10^{-19} coulombs
Electron rest mass	m	= 9.109×10^{-31} kg
	m	= 0.511 Mev = 0.00054859 amu
Proton rest mass	M	= 1.672×10^{-27} kg
	M	= 938.26 Mev = 1.007277 amu
Neutron rest mass	M	= 1.6748×10^{-27} kg
	M	= 939.550 Mev = 1.008665 amu
Alpha particle rest mass	M	= 4.0026 amu
	Pi = π	= 3.14159

Force constant- Coulomb's law	k	$= 8.988 \times 10^9 \text{ nm}^2/\text{coul}^2$
Rydberg constant-infinite mass	R_∞	$= 1.097373 \times 10^{-3} \text{ A}^{-1}$

Appendix 2

PERIODIC CHART OF THE ELEMENTS

Period	IA	IIA	IIIB	IVB	VB	VIB	VIIB	VIIIB			IB	IIB	IIIA	IVA	VA	VIA	VIIA	Inert Gases
1	1 H 1.00797																	2 He 4.0026
2	3 Li 6.939	4 Be 9.0122											5 B 10.811	6 C 12.01115	7 N 14.0067	8 O 15.9994	9 F 18.9984	10 Ne 20.183
3	11 Na 22.9898	12 Mg 24.312											13 Al 26.9815	14 Si 28.086	15 P 30.9738	16 S 32.064	17 Cl 35.453	18 Ar 39.948
4	19 K 39.102	20 Ca 40.08	21 Sc 44.956	22 Ti 47.90	23 V 50.942	24 Cr 51.996	25 Mn 54.9380	26 Fe 55.847	27 Co 58.9332	28 Ni 58.71	29 Cu 63.54	30 Zn 65.37	31 Ga 69.72	32 Ge 72.59	33 As 74.9216	34 Se 78.96	35 Br 79.909	36 Kr 83.80
5	37 Rb 85.47	38 Sr 87.62	39 Y 88.905	40 Zr 91.22	41 Nb 92.906	42 Mo 95.94	43 Tc (99)	44 Ru 101.07	45 Rh 102.905	46 Pd 106.4	47 Ag 107.870	48 Cd 112.40	49 In 114.82	50 Sn 118.69	51 Sb 121.75	52 Te 127.60	53 I 126.9044	54 Xe 131.30
6	55 Cs 132.905	56 Ba 137.34	57 *La 138.91	72 Hf 178.49	73 Ta 180.948	74 W 183.85	75 Re 186.2	76 Os 190.2	77 Ir 192.2	78 Pt 195.09	79 Au 196.967	80 Hg 200.59	81 Tl 204.37	82 Pb 207.19	83 Bi 208.980	84 Po (210)	85 At (210)	86 Rn (222)
7	87 Fr (223)	88 Ra (226)	89 †Ac (227)															

Series														
*Lanthanum Series	58 Ce 140.12	59 Pr 140.907	60 Nd 144.24	61 Pm (145)	62 Sm 150.35	63 Eu 151.96	64 Gd 157.25	65 Tb 158.924	66 Dy 162.50	67 Ho 164.930	68 Er 167.26	69 Tm 168.934	70 Yb 173.04	71 Lu 174.97
†Actinium Series	90 Th 232.038	91 Pa (231)	92 U 238.03	93 Np (237)	94 Pu (242)	95 Am (243)	96 Cm (247)	97 Bk (249)	98 Cf (251)	99 Es (254)	100 Fm (253)	101 Md (256)	102 No (253)	103 Lw (257)

The numbers in parentheses are the mass numbers of most stable or most common isotope.

A periodic table. Roman numerals designate groups; arabic, periods. Atomic numbers are listed above and left of element symbols; atomic weights, below.

Index